International Conference on Science of Friction 2010

(ICSF 2010)

Journal of Physics: Conference Series, Volume 258

Ise-Shima, Mie, Japan
13-18 September 2010

Editors:

Kouji Miura **Hiroshi Matsukawa**

ISBN: 978-1-61782-129-5
ISSN: 1742-6588

Printed from e-media with permission by:

Curran Associates, Inc.
57 Morehouse Lane
Red Hook, NY 12571

Some format issues inherent in the e-media version may also appear in this print version.

Printed by Curran Associates, Inc. (2010)

For permission requests, please contact the Institute of Physics
at the address below.

Institute of Physics
Dirac House, Temple Back
Bristol BS1 6BE UK

Phone: 44 1 17 929 7481
Fax: 44 1 17 920 0979

techtracking@iop.org

Additional copies of this publication are available from:

Curran Associates, Inc.
57 Morehouse Lane
Red Hook, NY 12571 USA
Phone: 845-758-0400
Fax: 845-758-2634
Email: curran@proceedings.com
Web: www.proceedings.com

TABLE OF CONTENTS

Preface

The second international conference on science of friction in Japan was held at Ise-Shima, Mie on 13–18 September 2010. The conference focused on the elementary process of friction phenomena from atomic and molecular scale view. Topics covered at the conference were:

- Superlubricity and friction
- Electronic and phononic contributions to friction
- Friction on the atomic and molecular scales
- van der Waals friction and Casimir force
- Molecular motor and friction
- Friction and adhesion in soft matter system
- Wear and crack on the nanocsale
- Theoretical studies on the atomic scale friction and energy dissipatin
- Friction and Chaos
- Mechanical properties of nanoscale contacts
- Friction of powder

The number of participants in the conference was approximately 85, registered from 8 countries. 40 oral and 16 poster talks were presented at the conference. This volume of *Journal of Physics: Conference Series* includes 19 papers devoted to the topics of friction.

The successful organization of the conference was made possible by the contribution of the members of the organizing Committee. The conference was made possible thanks to the financial support from Aichi University of Education, and moreover thanks to the approval societies of The Physical Society of Japan, The surface Science Society of Japan and The Japanese Society of Tribologists. The details of the conference are available on http://www.science-of-friction.com/2010/.

Finally we would like to thank the speakers for the high quality of their talks and all participants for coming to Ise-Shima, Japan and actively contributing to the conference.

Kouji Miura and Hiroshi Matsukawa
Editors

International Conference on Science of Friction 2010 (ICSF2010) IOP Publishing
Journal of Physics: Conference Series **258** (2010) 012001 doi:10.1088/1742-6596/258/1/012001

Normal and lateral Casimir force: Advances and prospects

G. L. Klimchitskaya

Department of Physics, North-West Technical University, Millionnaya Street 5, St.Petersburg, 191065, Russia
and
Institute for Theoretical Physics, Leipzig University, Postfach 100920, D-04009, Leipzig, Germany

E-mail: galina.klimchitskaya@itp.uni-leipzig.de

Abstract. We discuss recent experimental and theoretical results on the Casimir force between real material bodies made of different materials. Special attention is paid to calculations of the normal Casimir force acting perpendicular to the surface with the help of the Lifshitz theory taking into account the role of free charge carriers. Theoretical results for the thermal Casimir force acting between metallic, dielectric and semiconductor materials are presented and compared with available experimental data. Main attention is concentrated on the possibility to control the magnitude and sign of the Casimir force for applications in nanotechnology. In this respect we consider experiments on the optical modulation of the Casimir force between metal and semiconductor test bodies with laser light. Another option is the use of ferromagnetic materials, specifically, ferromagnetic dielectrics. Under some conditions this allows to get Casimir repulsion. The lateral Casimir force acting between sinusoidally corrugated surfaces can be considered as some kind of noncontact friction caused by zero-point oscillations of the electromagnetic field. Recent experiments and computations using the exact theory have demonstrated the role of diffraction-type effects in this phenomenon and the possibility to get asymmetric force profiles. Conclusion is made that the Casimir force may play important role in the operation of different devices on the nanoscale.

1. Introduction

Both the van der Waals and Casimir forces are quantum phenomena originating from the interaction of electromagnetic fluctuations with matter. One difference between them is that the van der Waals force is nonrelativistic, whereas the Casimir force takes full account of the relativistic retardation effects. Thus, in fact there exists just one force of a fluctuation nature. At shortest sepatations up to a few nanometers between the interacting bodies, where retardation is negligibly small, it is referred to as the *van der Waals* force. At larger separations it is called the *Casimir* force [1].

The Casimir force becomes significantly large if measured at separations below one micrometer. At separations below 100 nm it competes with characteristic electric forces acting between the constituting elements of microelectromechanical and nanoelectromechanical devices. The attractive Casimir force leads to pull-in and stiction [2] which may result in the collapse of a microdevice. That is why the Casimir force has attracted much recent attention in nanotechnology. On the one hand, it can be used as an actuating force [3] in place of electric forces. On the other hand, it plays the role of some kind friction in micrometer scales which cannot be diminished by means of lubrication. In this respect, the possibility of repulsive Casimir force attracts much public attention as the mean to cancel this friction.

It is commonly known that the unified description of both the van der Waals and Casimir forces is given by the Lifshitz theory [4, 5]. During more than fourty years after this theory was created it was lacking detailed experimental confirmation due to the absence of sufficiently precise measurements. During the last decade, however, many experiments measuring the Casimir force have been performed (see the recent review [6]). The comparison between the measurement data of the most precise experiments and computational results on the basis of the Lifshitz theory has led to a puzzle. It turned out that the Lifshitz theory is in agreement with the data only under a condition that for metallic test bodies the relaxation properties of conduction electrons are not accounted for, whereas for dielectrics charge carriers are completely discarded (controversial discussions concerning these results are reflected in Ref. [7]). Modern overview of both experimental and theoretical aspects of the Casimir force can be found in the monograph [8].

The present paper reviews several recent experimental and theoretical results in the Casimir effect which are of much importance for understanding of delicate physical mechanisms of fluctuation-induced forces and for their technological application in new generations of micro- and nanoelectromechanical systems. In Sec. 2 we briefly discuss the standard formalism of the Lifshitz theory, its generalization for the case of nonplanar geometries, and modifications aimed to resolve the above-mentioned puzzle. Section 3 is devoted to the comparison of theoretical results with the experimental data of three landmark experiments. These experiments were performed with metallic [9, 10, 11, 12], semiconductor [13, 14], and dielectric [15, 16] test bodies. In Sec. 4 the possibilities to control the magnitude and sign of the Casimir force by using materials with different electric and magnetic properties are considered. These include the variation of the density of free charge carriers with laser light, the use of three-layer systems and phase transitions, investigation of the Casimir force between ferromagnetic dielectrics etc. Special attention is devoted in Sec. 5 to the lateral Casimir force arising between sinusoidally corrugated surfaces, as predicted theoretically in Ref. [17] and experimentally demonstrated in Refs. [18, 19]. It was shown quite recently [20, 21] that this phenomenon has potential for applications in nanotechnology. Our conclusions and discussion are contained in Sec. 6.

2. Brief formulation of the Lifshitz theory

The Lifshitz theory [4, 5] expresses the free energy and pressure of the van der Waals and Casimir interaction in the configuration of two parallel semispaces (thick plates) separated with a gap of width a at temperature T in the following form:

$$\mathcal{F}(a,T) = \frac{k_B T}{2\pi} \sum_{l=0}^{\infty}{}' \Phi_E(\xi_l), \tag{1}$$

$$P(a,T) = -\frac{\partial \mathcal{F}(a,T)}{\partial a} = -\frac{k_B T}{\pi} \sum_{l=0}^{\infty}{}' \Phi_P(\xi_l).$$

Here, k_B is the Boltzmann constant, $\xi_l = 2\pi k_B T l/\hbar$ with $l = 0, 1, 2, \ldots$ are the Matsubara frequencies, and the terms with $l = 0$ in the primed sums are multiplied by $1/2$. The functions Φ_E and Φ_P are expressed in terms of reflection coefficients on the two semispaces $r_\alpha^{(1)}$ and $r_\alpha^{(2)}$ for two independent polarizations of the electromagnetic field, transverse magnetic ($\alpha = \mathrm{TM}$) and transverse electric ($\alpha = \mathrm{TE}$),

$$\Phi_E(x) = \int_0^\infty k_\perp dk_\perp \sum_\alpha \ln\left[1 - r_\alpha^{(1)}(ix, k_\perp) r_\alpha^{(2)}(ix, k_\perp)\, e^{-2aq}\right],$$

$$\Phi_P(x) = \int_0^\infty k_\perp dk_\perp q \sum_\alpha \left[\frac{e^{2aq}}{r_\alpha^{(1)}(ix,k_\perp)r_\alpha^{(2)}(ix,k_\perp)} - 1\right]^{-1}, \qquad (2)$$

where

$$q \equiv q(ix,k_\perp) = \sqrt{k_\perp^2 + x^2/c^2}, \qquad (3)$$

and $\boldsymbol{k}_\perp = (k_x, k_y)$ is the projection of a wave vector on the plane of plates. The reflection coefficients on homogeneous media described by the frequency-dependent dielectric permittivity $\varepsilon^{(n)}(\omega)$ and magnetic permeability $\mu^{(n)}(\omega)$ $(n = 1,\, 2)$ are given by

$$r_{\rm TM}^{(n)}(ix,k_\perp) = \frac{\varepsilon^{(n)}(ix)q - k^{(n)}}{\varepsilon^{(n)}(ix)q + k^{(n)}}, \qquad r_{\rm TE}^{(n)}(ix,k_\perp) = \frac{\mu^{(n)}(ix)q - k^{(n)}}{\mu^{(n)}(ix)q + k^{(n)}}, \qquad (4)$$

where

$$k^{(n)} \equiv k^{(n)}(ix,k_\perp) = \sqrt{k_\perp^2 + \varepsilon^{(n)}(ix)\mu^{(n)}(ix)\frac{x^2}{c^2}}. \qquad (5)$$

The Lifshitz-type formulas similar to (1) hold also for the Casimir-Polder free energy and force acting between an atom with dynamic electric polarizability $\alpha(\omega)$ and magnetic susceptibility $\beta(\omega)$ and a wall decribed by $\varepsilon(\omega)$ and $\mu(\omega)$

$$\mathcal{F}_A(a,T) = -k_BT\sum_{l=0}^{\infty}{}' \int_0^\infty k_\perp dk_\perp q_l \Phi_A(\xi_l,k_\perp), \qquad (6)$$
$$F_A(a,T) = -k_BT\sum_{l=0}^{\infty}{}' \int_0^\infty k_\perp dk_\perp q_l^2 \Phi_A(\xi_l,k_\perp),$$

where $q_l = q(i\xi_l,k_\perp)$. The function $\Phi_A(\xi_l,k_\perp)$ is defined as [22]

$$\Phi_A(x,k_\perp) = e^{-2aq}\left\{2\left[\alpha(ix)\,r_{\rm TM}(ix,k_\perp) + \beta(ix)\,r_{\rm TE}(ix,k_\perp)\right]\right. \qquad (7)$$
$$\left. - \frac{x^2}{q^2c^2}\left[\alpha(ix)+\beta(ix)\right]\left[r_{\rm TM}(ix,k_\perp) + r_{\rm TE}(ix,k_\perp)\right]\right\}.$$

Due to the presence of only one wall, the upper index on the reflection coefficients is omitted.

Using the so-called *proximity force approximation* (PFA), the Casimir force acting between a sphere and a plate can be expressed as

$$F(a,T) = 2\pi R\mathcal{F}(a,T), \qquad (8)$$

where the free energy per unit area for two parallel semispaces is defined in Eq. (1). Under the condition that $a \ll R$, where R is the sphere radius and a is the closest separation between the surfaces of a sphere and a plate, the relative error of the approximate Eq. (8) is less than a/R [6, 8]. Equation (8) is widely used for the comparison of the experimental data obtained from the measurements of the Casimir force between a sphere and a plate with the predictions of the Lifshitz theory.

The original formulation of the Lifshitz theory for two parallel semispaces separated with a gap can be simply generalized for an arbitrary number of planar layers of different thickness made of arbitrary materials [23, 24, 25]. However, the obtaining of exact formulas for the Casimir free energy and force between bodies with nonplanar surfaces was challenging during the last few decades. The problem was solved in 2006 when several equivalent Lifshitz-type formulas were obtained [26, 27, 28, 29] using the scattering theory and the formalism of functional determinants. Specifically, in one of the developed formalisms the Casimir free energy for two separated bodies of arbitrary shape labeled by the indices (1) and (2) can be presented in the form [30]

$$\mathcal{F}^{\rm exact}(a,T) = k_BT\sum_{l=0}^{\infty}{}' \ln\det\left[1-\mathcal{T}^{(1)}\mathcal{G}^{(0)}_{\xi_l;(1,2)}\mathcal{T}^{(2)}\mathcal{G}^{(0)}_{\xi_l;(2,1)}\right]. \qquad (9)$$

Here, $\mathcal{G}^{(0)}_{\xi_l;(1,2)}$ is the operator for the free space Green's function with the matrix elements $\langle \boldsymbol{r}|\mathcal{G}^{(0)}_{\xi_l;(1,2)}|\boldsymbol{r}'\rangle$, where $\boldsymbol{r}$ belongs to the body (1) and $\boldsymbol{r}'$ belongs to the body (2). $\mathcal{T}^{(1)}$ $[\mathcal{T}^{(2)}]$ is the operator of T matrix for the body (1) [body (2)]. Using such representations, it was shown that for an ideal metal sphere above an ideal metal plate at room temperature the exact results for the Casimir force and for the thermal correction to it coincide with respective results obtained using the PFA in the zeroth order of the small parameter a/R [31].

Calculation of the Casimir force between real material bodies using the Lifshitz theory requires the availability of the respective $\varepsilon(\omega)$ and $\mu(\omega)$. We reserve the discussion of $\mu(\omega)$ for Sec. 4 and briefly consider here the dielectric permittivity for dielectrics and metals. It is common knowledge that all condensed bodies are either dielectrics or metals depending on the behavior of their dc conductivity $\sigma_0(T)$ when temperature vanishes [32, 33]. For dielectrics $\sigma_0(0) = 0$ whereas for metals $\sigma_0(0) \neq 0$. In this regard semiconductors are also either dielectrics or metals (specifically, intrinsic semiconductors are dielectrics, whereas doped semiconductors with dopant concentration above the critical value are metals). In the application of the Lifshitz theory to dielectric materials the small dc conductivity arising at $T \neq 0$ is usually neglected. In so doing the dielectric permittivities of dielectric plates are determined by the core electrons [34]

$$\varepsilon_c^{(n)}(\omega) = 1 + \sum_{j=1}^{K}\frac{g_j^{(n)}}{\omega_j^{(n)2} - \omega^2 - i\gamma_j^{(n)}\omega}, \qquad (10)$$

where $\omega_j^{(n)} \neq 0$ are the oscillator frequencies, $g_j^{(n)}$ are the oscillator strengths, $\gamma_j^{(n)}$ are the damping parameters, and K is the number of oscillators. Note that the parameters of oscillators in Eq. (10) can be temperature-dependent. Thus, in addition to the explicit dependence on the temperature in Eqs. (1) and (6) through the multiple T and through the Matsubara frequencies, there is an implicit dependence through the dielectric permittivity. For Si such an implicit dependence was investigated in Ref. [35].

However, if one takes into account the existence of free charge carriers at $T \neq 0$, the dielectric permittivity of dielectric plates can be represented as [36]

$$\varepsilon_d^{(n)}(\omega) = \varepsilon_c^{(n)}(\omega) + i\frac{4\pi\sigma_0^{(n)}(T)}{\omega}. \qquad (11)$$

It has been shown [37, 38] that the Casimir entropy calculated using the free energy (1) combined with the dielectric permittivity (10) satisfies an important thermodynamic constraint

$$S_c(a,T) = -\frac{\partial\mathcal{F}_c(a,T)}{\partial T} \to 0 \qquad (12)$$

when $T \to 0$ (the Nernst heat theorem). If, however, the dc conductivity of dielectric material is taken into account [i.e., the dielectric permittivity (11) is used] we get [37, 38]

$$S_d(a,0) = \frac{k_B}{16\pi a^2}\left[\zeta(3) - \mathrm{Li}_3(r_0^{(1)}r_0^{(2)})\right] > 0, \qquad (13)$$

where

$$r_0^{(n)} = \frac{\varepsilon_c^{(n)}(0) - 1}{\varepsilon_c^{(n)}(0) + 1} < 1, \tag{14}$$

$\zeta(z)$ is the Riemann zeta function, and $\mathrm{Li}_n(z)$ is the polylogarithm function. Keeping in mind that real dielectric materials possess dc conductivity, one arrives to the conclusion that for two parallel dielectric semispaces the Lifshitz theory violates the Nernst teorem.

A similar difficulty arises in the application of the Lifshitz theory to real metals. It is customary to describe the dielectric properties of metallic plates by means of the Drude model

$$\varepsilon_D^{(n)}(\omega) = 1 - \frac{\omega_p^{(n)2}}{\omega[\omega + i\gamma^{(n)}(T)]}, \tag{15}$$

where $\omega_p^{(n)}$ are the plasma frequencies and $\gamma^{(n)}(T)$ are the relaxation parameters. This model is well applicable in the wide frequency region from quasistatic frequencies (the normal skin effect) to infrared optics. It was shown [6, 8, 39, 40] that for perfect crystal lattices [where $\gamma^{(n)}(T) \to 0$ as T^2] the Casimir entropy calculated using the free energy (1) combined with the Drude model (15), violates the Nernst heat theorem

$$S_D(a, 0) = -\frac{k_B\zeta(3)}{16\pi a^2}\left[1 - \frac{2c}{a}\frac{\omega_p^{(1)} + \omega_p^{(2)}}{\omega_p^{(1)}\omega_p^{(2)}} + \frac{3c^2}{a^2}\left(\frac{\omega_p^{(1)} + \omega_p^{(2)}}{\omega_p^{(1)}\omega_p^{(2)}}\right)^2 - \ldots\right] < 0. \tag{16}$$

Notice that for Drude metals with impurities there is a nonzero residual relaxation $\gamma_{\mathrm{res}}^{(n)} = \gamma^{(n)}(0)$. In this case the Casimir entropy jumps abruptly to zero at $T < 10^{-4}\,$K starting from a negative value (16), so that the Nernst theorem is formully satisfied [41, 42, 43]. This, however, does not solve the problem. The point is that a perfect crystal lattice is a truly equilibrium system with a nondegenerate dynamical state of the lowest energy. Thus, according to quantum statistical physics, the Casimir entropy calculated for a perfect crystal lattice must be equal to zero [44].

It is quite another matter when the nondissipative plasma model

$$\varepsilon_p^{(n)}(\omega) = 1 - \frac{\omega_p^{(n)2}}{\omega^2}, \tag{17}$$

is used instead of (15) to describe metallic plates. In this case the Nernst heat theorem is satisfied [6, 8, 39, 40], i.e.

$$S_p(a, T) = -\frac{\partial \mathcal{F}_p(a, T)}{\partial T} \to 0 \tag{18}$$

when $T \to 0$. It was proposed [6, 8, 45] that the inclusion of relaxation properties of free charge carriers in the model of dielectric response violates thermal equilibrium which is the basic applicability condition of the Lifshitz theory. This is because the drift electric current leads to a violation of time-reversal symmetry and requires the introduction of the unidirectional flux of heat from the Casimir plates to the heat reservoir to preserve the temperature constant. Consensus on the problem of why the Lifshitz theory is thermodynamically inconsistent in the presence of free charge carriers is not yet achieved.

To complete a brief survey of the Lifshitz theory, we mention recent attempts to modify the TM reflection coefficients by taking into account the effect of screening [46, 47]. This approach includes both the drift and diffusion currents of free charge carriers. Eventually the modified

TM reflection coefficient is obtained through the use of Boltzmann transport equation and takes the form [47]

$$r_{\rm TM}^{(n)\rm mod}(ix,k_\perp) = \frac{\varepsilon^{(n)}(ix)q - k^{(n)} - \frac{k_\perp^2}{\eta^{(n)}(ix)}\frac{\varepsilon^{(n)}(ix)-\varepsilon_c^{(n)}(ix)}{\varepsilon_c^{(n)}(ix)}}{\varepsilon^{(n)}(ix)q + k^{(n)} + \frac{k_\perp^2}{\eta^{(n)}(ix)}\frac{\varepsilon^{(n)}(ix)-\varepsilon_c^{(n)}(ix)}{\varepsilon_c^{(n)}(ix)}}. \qquad (19)$$

Here, the dielectric permittivities of the plates are given by

$$\varepsilon^{(n)}(ix) = \varepsilon_c^{(n)}(ix) + \frac{\omega_p^{(n)\,2}}{x(x+\gamma^{(n)})}, \qquad (20)$$

where permittivities of the bound core electrons $\varepsilon_c^{(n)}$ are defined in Eq. (10). The quantity $\eta^{(n)}(ix)$ is given by

$$\eta^{(n)}(ix) = \left[k_\perp^2 + \kappa^{(n)2}\,\frac{\varepsilon_c^{(n)}(0)}{\varepsilon_c^{(n)}(ix)}\,\frac{\varepsilon^{(n)}(ix)}{\varepsilon^{(n)}(ix)-\varepsilon_c^{(n)}(ix)}\right]^{1/2}, \qquad (21)$$

where $1/\kappa^{(n)}$ are the screening lengths of plate materials [48] defined differently for dielectrics and metals (the Debye-Hückel and Tomas-Fermi screening lengths when the charge carriers are described by the Maxwell-Boltzmann and Fermi-Dirac statistics, respectively). The TE reflection coefficient in the proposed approach remains unchanged.

It was shown [45, 49, 50, 51], however, that the modification of the TM reflection coefficient in accordance with Eq. (19) also leads to the violation of the Nernst theorem for all dielectric materials whose charge carrier density does not vanish when $T \to 0$ and $\sigma_0(T)$ vanishes due to the vanishing mobility (for instance, for glasses with ionic conductivity [52] or doped semiconductors with dopant concentration below the critical value [53]). For metals with perfect crystal lattices the Lifshitz theory with the modified TM reflection coefficient violates the Nernst theorem in the same way as when the standard Drude model approach is used [50, 54, 55]. These conclusions were disputed in Refs. [56, 57], but the argumentation used was shown to be inadequate in Refs. [7, 50]. Therefore, the problem of consistency of the Lifshitz theory with thermodynamics for real materials remains unsolved.

3. Comparison with three landmark experiments

There are about 25 experiments on measuring the Casimir force in the modern stage of the work on the subject (see the review [6] for more details). In almost all experiments at least an attempt of comparing the measurement data with theory was undertaken. In this section, we consider only three experiments each of which, when carefully compared with the Lifshitz theory, leads to unexpected conclusions.

3.1. Measuring the Casimir pressure using a micromachined oscillator

The first of this landmark experiments was performed by R. S. Decca and his co-workers. It was repeated three times, in 2003 [9], 2005 [10], and in 2007 [11, 12] with major improvements in sample preparation and in the experimental precision in each repetition. The discussion below is related to the most precise experiment of 2007 [11, 12], where the gradient of the Casimir force between an Au-coated sphere of $R = 151.3\,\mu$m radius and a plate of micromachined oscillator was measured at $T = 300\,$K in the dynamic regime (see Fig. 1 for the schematic diagram of a setup). In fact it was the shift of the resonant frequency of a micromachined oscillator that was

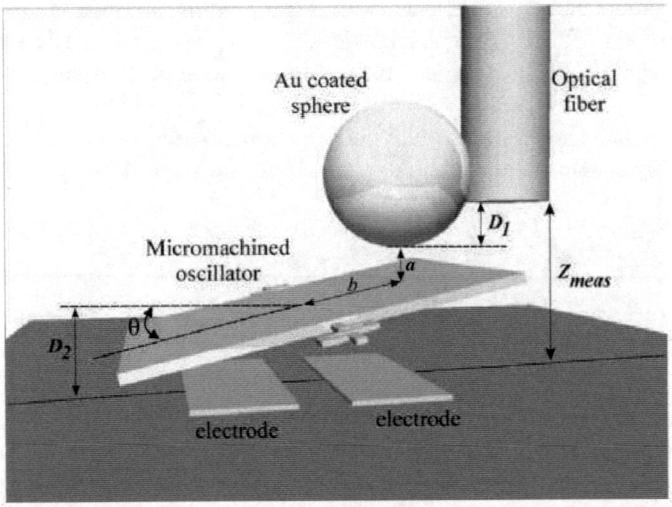

Figure 1. Schematic diagram of the setup for measuring the gradient of the Casimir force between a sphere and a plate (which is also an indirect measurement of the Casimir pressure between two parallel plates) using a micromachined oscillator.

measured. This shift is proportional to the gradient of the Casimir force between a sphere and a plate. Considering the negative derivative with respect to a of both sides of Eq. (8), one obtains

$$P(a,T) \equiv -\frac{\partial \mathcal{F}(a,T)}{\partial a} = -\frac{1}{2\pi R}\frac{\partial F(a,T)}{\partial a}. \tag{22}$$

Thus, the use of the PFA allows interpretation of the experiment by Decca et al. as an indirect measurement of the Casimir pressure between two parallel plates. In this experiment all errors were determined at a 95% confidence level, and the random errors were made much less than the systematic errors, so that only the latter determined the total experimental error. As a result, the total relative experimental error of the Casimir pressure varied from 0.19% at $a = 162\,$nm to 0.9% at $a = 400\,$nm and to 9.0% at the largest separation $a = 746\,$nm. This includes the error of the PFA which was used to recalculate the experimental data for the gradient of the Casimir force into the Casimir pressure. In the most conservative way the error of the PFA was taken to be equal to a/R (specially performed experiment [58] and theoretical computations [59, 60, 61] for a sphere and a plate made of real metals described by the plasma and Drude models show that this error is in fact smaller than a/R).

Computations of the Casimir pressure between Au plates was performed by Eq. (1) with $\mu^{(n)} = 1$ using the tabulated optical data [36] for the complex index of refraction $n(\omega) = \mathrm{Re}\,n(\omega) + i\mathrm{Re}\,n(\omega)$. For this purpose the imaginary part of the dielectric permittivity of Au was found as $\mathrm{Im}\,\varepsilon(\omega) = 2\mathrm{Re}\,n(\omega)\mathrm{Im}\,n(\omega)$ at frequencies where the optical data are available and extrapolated to lower frequencies by means of the Drude model (15) with the parameters $\omega_p = 9.0\,$eV, $\gamma = 0.035\,$eV [62]. Then $\varepsilon(i\xi)$ for Au was obtained using the Kramers-Kronig relation

$$\varepsilon(i\xi) = 1 + \frac{2}{\pi}\int_0^\infty \frac{\omega\,\mathrm{Im}\,\varepsilon(\omega)}{\xi^2+\omega^2}\,d\omega. \tag{23}$$

This calculational approach corresponds to the use of the Drude model, as described in Sec. 2, but takes into account the role of core electrons. Keeping in mind that problems discussed in Sec. 2 originate from the zero-frequency term of the Lifshitz formula, one might expect that related difficulties should appear here as well. For the comparisom with the experimental data, the surface roughness on both the sphere and the plate was studied with the help of an atomic force microscope and taken into account in a nonmultiplicative way by means of geometrical averaging [6, 8].

Another approach to calculate the Casimir pressure between real metals with account of core electrons uses the generalized plasma-like model for the permittivity

$$\varepsilon(\omega) = \varepsilon_c(\omega) - \frac{\omega_p^2}{\omega^2}, \tag{24}$$

where $\varepsilon_c(\omega)$ is defined in Eq. (10) and the parameters of oscillators for Au with $K = 6$ were determined in Ref. [12]. The surface roughness is taken into account as described above.

Now we present a typical comparison between the experimental data and theory. In Fig. 2(a,b) the mean Casimir pressure together with the absolute experimental errors in the separation and pressure measured at different separations is shown as crosses. In Fig. 2(a) the theoretical Casimir pressure computed using the generalized plasma-like model and the optical data of Ref. [36] extrapolated by the Drude model is shown by the light-gray and dark-gray bands, respectively. The width of the bands indicates the total theoretical error. As can be seen in Fig. 2(a), the plasma-like model is consistent with the data, whereas a theory using the extrapolation to low frequencies by means of the Drude model is excluded by the data at a 95% confidence level. As was noted in the literature [63, 64], Au films prepared using different technology may possess different sets of optical data leading to the values of ω_p varying from 6.85 to 9.0 eV. In Fig. 2(b) the theoretical Casimir pressure as a function of separation computed with all sets of optical data available is shown as the dark-gray band. It is seen that the use of any alternative value of ω_p makes the disagreement between the Drude model approach and the data even more evident. One more recent experiment [65] incorporated the determination of the Casimir pressure by means of a micromachined oscillator with simultaneous measurement of the optical data of Au films. It was shown that while the Casimir pressures coincide with those measured previously in Refs. [11, 12], the optical data closely follow those tabulated in Ref. [36].

Another method for comparing theory with experiment considers the differences between theoretical and mean experimental Casimir pressures, $P^{\rm theor}(a_i) - \bar{P}^{\rm expt}(a_i)$, and the 95%

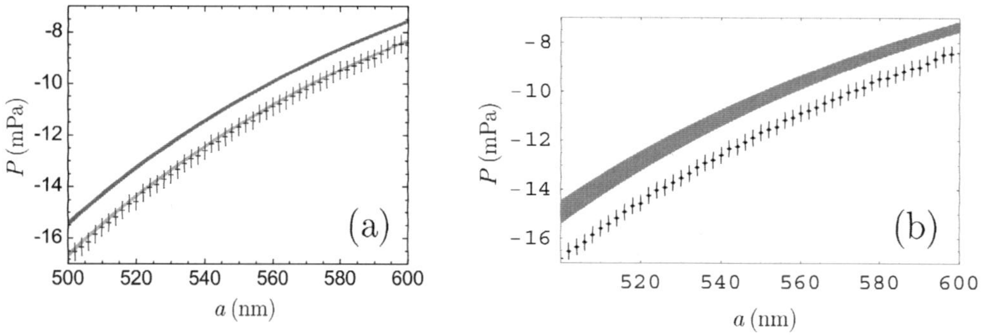

Figure 2. The measured mean Casimir pressure together with the absolute errors in the separation and pressure versus separation is shown as crosses. Errors are calculated at a 95% confidence level. (a) The theoretical Casimir pressure computed using the generalized plasma-like model and the optical data extrapolated by the Drude model is shown by the light-gray and dark-gray bands, respectively. (b) The theoretical Casimir pressure computed using different sets of optical data available in the literature versus separation is shown as the dark-gray band.

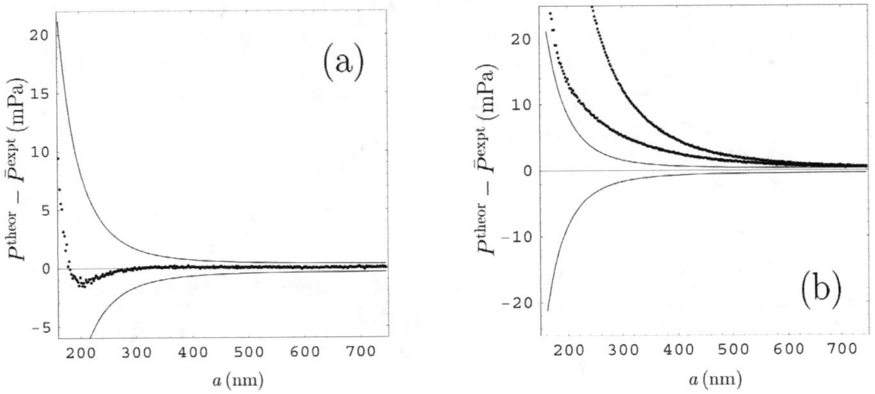

Figure 3. Differences between the theoretical Casimir pressures, computed by means of (a) the generalized plasma-like model and (b) the Drude model approach, and mean experimental Casimir pressures at different separations are shown as dots. The lower and upper sets of dots for the Drude model approach correspond to the use of maximum and minimum values of the plasma frequency available in the literature. The solid lines indicate the borders of the confidence intervals found at a 95% confidence level.

confidence interval $[-\Xi_P(a), \Xi_P(a)]$ to which 95% of these differences must belong if the theory is consistent with the data. In Fig. 3(a), the differences of the theoretically computed pressures using the generalized plasma-like model and mean experimental Casimir pressures are shown as dots. The solid lines are formed by the ends of the confidence intervals. As can be seen in Fig. 3(a), the Lifshitz theory using the generalized plasma-like model is consistent with the data. A similar comparison is made with the theoretical Casimir pressures computed using the tabulated optical data extrapolated by the Drude model. In Fig. 3(b) the differences $P^{\rm theor}(a_i) - \bar{P}^{\rm expt}(a_i)$ are indicated as two sets of dots computed using $\omega_p = 6.82\,$eV (the upper set) and $\omega_p = 9.0\,$eV (the lower set). The pressure differences for all other values of the plasma frequency of Au considered in Ref. [64] are between the two sets of dots in Fig. 3(b). Thus, theory using the Drude model is experimentally excluded over the entire measurement range from 162 to 746 nm at a 95% confidence level. Note that within a more narrow range of separations from 210 to 620 nm the theoretical approach using the Drude model is excluded by the data at a 99.9% confidence level [6, 8, 12]. The same data exclude [66] the Lifshitz theory at zero temperature combined with the Drude model. This exclusion is made at a 70% confidence level with the optical data of Ref. [36] and at 95% confidence level with the alternative optical data.

The experimental data of Ref. [11, 12] have been also compared with theoretical predictions obtained using the modified TM reflection coefficient (19). In so doing it was supposed that electrons in Au are described by Fermi-Dirac statistics. Almost the same computational results for the Casimir pressure were obtained [50, 54, 55], as by using the Drude model. Thus, the Lifshitz theory with the modified TM reflection coefficient is excluded by the experiment of Refs. [11, 12].

3.2. Optical modulation of the Casimir force

The second landmark experiment was performed by U. Mohideen and his co-workers [13, 14]. In this experiment, an atomic force microscope was employed to measure the change in the Casimir force between an Au-coated sphere ($R = 98.9\,\mu$m) and a Si membrane at $T = 300\,$K in the presence and in the absence of incident laser light on a membrane (see schematic of the experimental setup in Fig. 4). The experiment was repeated three times with three different

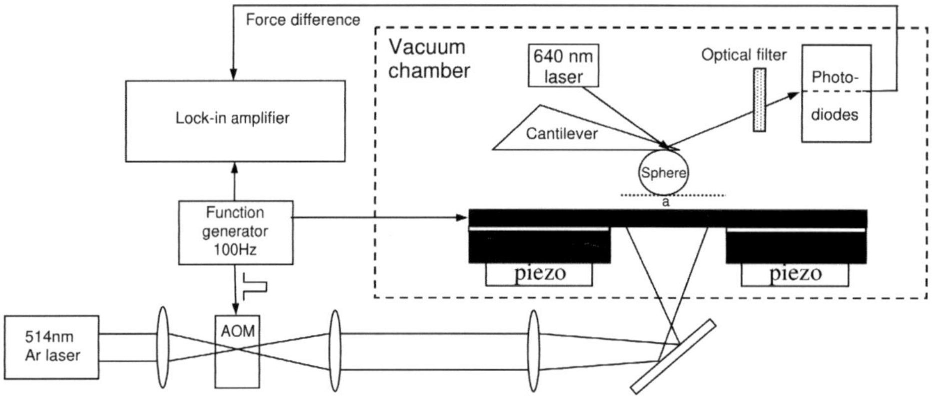

Figure 4. Schematic of the experimental setup on measuring the difference Casimir force between Au sphere and Si plate illuminated with laser pulses.

absorbed laser powers. The values of the maximum and minimum absorbed powers were equal to $P_a = 9.3\,\mathrm{mW}$ and $P_b = 4.7\,\mathrm{mW}$. In the absence of laser light, the charge carrier density of high-resistivity Si was $n_h = 5\times10^{14}\,\mathrm{cm}^{-3}$. In the presence of light, the charge carrier densities related to the maximum and minimum absorbed powers took the values $n_a = (2.1\pm0.4)\times10^{19}\,\mathrm{cm}^{-3}$ and $n_b = (1.4\pm0.3)\times10^{19}\,\mathrm{cm}^{-3}$ (the errors are indicated at a 95% confidence level [14]).

The difference of the theoretical Casimir forces in the presence and in the absence of laser light,

$$F_{\mathrm{diff}}^{\mathrm{theor}}(a,T) = F_l^{\mathrm{theor}}(a,T) - F^{\mathrm{theor}}(a,T), \tag{25}$$

was calculated using Eq. (1) with $\mu^{(n)} = 1$ and Eq. (8), taking into account the surface roughness by means of geometrical averaging. The dielectric permittivity of the Au-coated sphere $\varepsilon^{(1)}(i\xi) \equiv \varepsilon_{\mathrm{Au}}(i\xi)$ is described by Eq. (24). The dielectric permittivity of the Si plate $\varepsilon^{(2)}(i\xi) \equiv \varepsilon_{\mathrm{Si}}(i\xi)$ depends on the concentration of charge carriers. In the absence of charge carriers (infinitely high resistivity) it was computed by using the tabulated optical data [36] and the Kramers-Kronig relation. The obtained results, $\varepsilon_{\mathrm{Si},0}(i\xi)$, are shown by the dashed line in Fig. 5(a). The dielectric permittivity of high-resistivity plate used in the experiment in the absence of laser light, $\varepsilon_{\mathrm{Si},h}(i\xi)$, can be presented as the sum of $\varepsilon_{\mathrm{Si},0}(i\xi)$ and the second term on the right-hand side of Eq. (11) taking the dc conductivity into account. In Fig. 5(a) the quantity $\varepsilon_{\mathrm{Si},h}(i\xi)$ is shown by the dotted line. Similarly, the dielectric permittivity of low-resistivity Si with different absorbed powers, $\varepsilon_{\mathrm{Si},a,b}(i\xi)$, are shown by the solid lines a and b taking into account both electrons and holes [67, 68, 69].

Now we are in a position to compare the experimental data with the predictions of the Lifshitz theory. In Fig. 5(b) the mean experimental differences, $\bar{F}_{\mathrm{diff}}^{\mathrm{expt}}$, of the Casimir forces in the presence and in the absence of laser light for the absorbed power P_b are shown as crosses. The arms of the crosses represent total experimental errors in the measurement of the difference force and separation distance, determined at a 95% confidence level. The theoretical differences of the Casimir force (25) were first computed using the dielectric permittivity $\varepsilon_{\mathrm{Si},h}(i\xi)$ in the absence of light which takes the dc conductivity of a Si plate into account. In the presence of light the permittivity $\varepsilon_{\mathrm{Si},b}(i\xi)$ has been used. The computational results are presented by the dashed line in Fig. 5(b). As can be seen in the figure, the prediction of the Lifshitz theory taking into account the dc conductivity of high-resistivity Si is experimentally excluded. Then the theoretical differences (25) were computed using the dielectric permittivity $\varepsilon_{\mathrm{Si},0}(i\xi)$ in the

Journal of Physics: Conference Series **258** (2010) 012001 doi:10.1088/1742-6596/258/1/012001

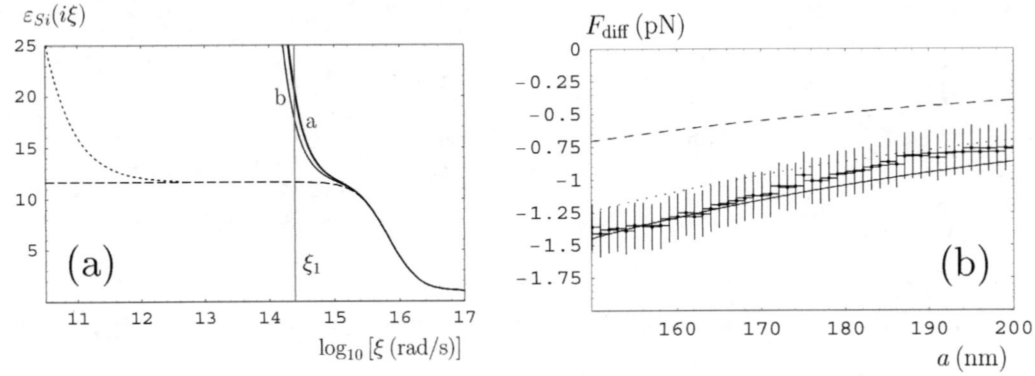

Figure 5. (a) The dielectric permittivity of the Si membrane along the imaginary frequency axis for Si with different concentration of charge carriers (see text for further discussion). The vertical gray line indicates the first Matsubara frequency at $T = 300\,$K. (b) The experimental differences in the Casimir force with their experimental errors are shown as crosses (the absorbed power is equal to 4.7 mW). The solid and dotted lines represent the theoretical differences computed at $T = 300\,$K using the model with a finite static permittivity of high-resistivity Si, but different models for Si in the presence of light (see text for further discussion). The dashed line represents the theoretical differences computed including the dc conductivity of high-resistivity Si.

dark phase which disregards the dc conductivity of Si. In the presence of light the permittivity $\varepsilon_{\text{Si},b}(i\xi)$ was again used. In this case the computational results are shown by the solid line in Fig. 5(b) which is consistent with the data.

Note that the dielectric permittivity $\varepsilon_{\text{Si},b}(i\xi)$ used in the presence of light describes charge carriers in the low-resistivity Si by means of the Drude model. If to take into account that n_b is larger than the critical value of charge carrier density indicating the dielectric to metal phase transition, it would be reasonable to use the description by means of the generalized plasma-like model in the presence of light (see Sec. 2). The computational results obtained within this approach [70] are presented by the dotted line in Fig. 5(b). It is seen that the dotted line is consistent with the data, as well as the solid line. Thus, this experiment is not of sufficient precision to discriminate between the description of charge carriers in metallic phase by means of the Drude and plasma models (this was done by the experiment of Decca et al. considered in Sec. 3.1). The main new and unexpected result of the experiment by Mohideen et al. is that the inclusion of the dc conductivity in a dielectric phase (high-resistivity Si) results in the contradiction of the Lifshitz theory with the data. Because of this, phenomenologically one may conclude [38] that in applications of the Lifshitz theory to dielectric materials the dc conductivity should be disregarded. Note that the contribution of dc conductivity to the dielectric permittivity of dielectrics was also disregarded when studying nonequilibrium thermal fluctuations of the electric field in the presence of a dc current [71].

The experimental data of the optical modulation experiment was compared with the modified Lifshitz theory using the TM reflection coefficient (19) proposed in Refs. [46, 47]. It was found that predictions of the modified Lifshitz theory for the Si plate in a dielectric phase cannot be excluded at a 95% confidence level. Because of this, the comparison has been made at a 70% confidence level [45]. In Fig. 6(a) the experimental data for the measured difference Casimir force in the presence and in the absence of light are shown as crosses. When compared to Fig. 5(b), the arms of the crosses are made relatively shorter, as required by a lower confidence level. The prediction of the modified Lifshitz theory for the absorbed power P_b is shown as a band bounded by the two dashed lines. The width of this band is found from the theoretical error determined mostly by the error in n_b (at a 70% confidence level this error is equal to

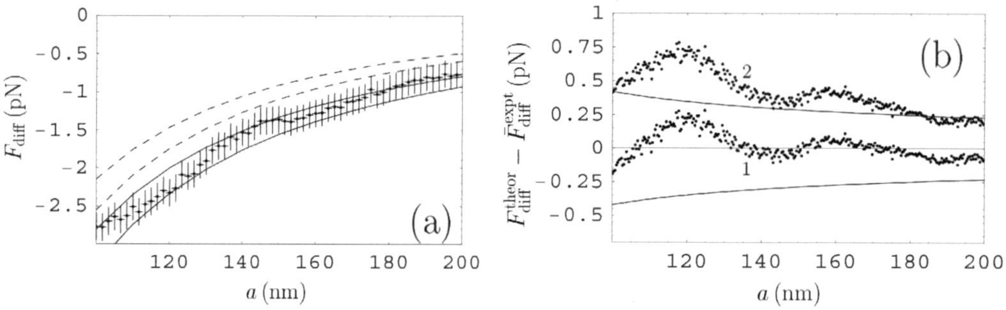

Figure 6. (a) The measured force differences are shown as crosses versus separation a. The two theoretical bands lie between the solid and dashed lines (see text for details). (b) Theoretical minus mean experimental differences of the Casimir force versus separation are shown as dots. For the dots labeled 1, the theoretical results are computed with a finite static dielectric permittivity of high resistivity Si in the absence of laser pulse and using the plasma model to describe the charge carriers when the laser pulse is on. For dots labeled 2, the theoretical results are computed in the framework of the modified Lifshitz theory [46, 47]. The solid lines indicate 70% confidence intervals including all experimental and theoretical errors.

$0.22 \times 10^{19}\,\mathrm{cm}^{-3}$ to compare with $0.3 \times 10^{19}\,\mathrm{cm}^{-3}$ at a 95% confidence level). The prediction of the standard Lifshitz theory with dc conductivity of Si in the absence of light disregarded is shown by the band bounded by two solid lines. The width of this band also corresponds to an error in n_b determined at a 70% confidence level. As can be seen in Fig. 6(a), the predictions of the modified Lifshitz theory [46, 47] are excluded by the data at a 70% confidence level, whereas the standard Lifshitz theory with disregarded dc conductivity of dielectric Si is consistent with the data.

The same conclusion can be obtained when using another approach to the comparison between experiment and theory similar to that in Fig. 3. In Fig. 6(b) the differences between theoretical and mean experimental difference Casimir forces, $F_{\rm diff}^{\rm theor} - \bar{F}_{\rm diff}^{\rm expt}$, at different separations are plotted as the two sets of dots. The set labeled 1 is computed using the standard Lifshitz theory with the dc conductivity of dielectric Si disregarded. The set labeled 2 is computed using the modified Lifshitz theory. The two solid lines indicate the borders of a 70% confidence intervals determined with account of all experimental and theoretical errors. It is seen that the standard Lifshitz theory with the dc conductivity disregarded is consistent with the data, whereas the modified Lifshitz theory [46, 47] is excluded by the measurement results of the optical modulation experiment at a 70% confidence level.

3.3. Measurements of the thermal Casimir-Polder force

The third landmark experiment is on measuring the Casimir-Polder force between ground state ^{87}Rb atoms and a fused silica plate. It was performed by Cornell and his co-workers [15]. Atoms of ^{87}Rb belonged to the Bose-Einstein condensate separated by a distance of a few micrometers from a wall. The experiment was of dynamic-type. The dipole oscillations in the direction perpendicular to the plate with the frequency ω_0 were excited in a condensate, and the relative shift of this frequency

$$\gamma_z = \frac{|\omega_0 - \omega_z|}{\omega_0} \tag{26}$$

due to the Casimir-Polder interaction was measured precisely. This shift can be recalculated [72] into the Casimir-Polder force (6) with $\mu^{(n)} = 1$, $\beta(i\xi) = 0$ leading to its indirect measurement.

This experiment was repeated three times, once in thermal equilibrium, when the temperature of the plate was equal to the environment temperature, $T_p = T_e = 310\,$K, and two times with $T_p > T_e$ (in the latter case there are additions to the Casimir-Polder force (6) calculated in Ref. [73]). The relative frequency shift (26) was measured at several atom-plate separations in the region from 7 to 11 μm. Keeping in mind that the comparison of the frequency shift with the Casimir-Polder force includes an averaging over the condensate cloud, the role of atom-plate separation is in fact played by the distance between the plate and the center of mass of the condensate.

In Fig. 7(a) the experimental data for the relative frequency shift γ_z in thermal equilibrium at separations below 10 μm are shown as crosses [15]. The arms of the crosses indicate total experimental errors in the measurement of γ_z and separation distances at a 66% confidence level found at each experimental point separately using the analysis of Ref. [74]. To compare the experimental data with theory, in Ref. [15] the relative frequency shift (26) was computed using the Lifshitz formula (6) for the Casimir-Polder force and an averaging over the condensate cloud. In so doing the static value for the polarizability of Rb atom $\alpha(0) = 4.73\times10^{-23}\,\text{cm}^3$ can be used. The dielectric permittivity of fused silica $\varepsilon(i\xi_l)$, as a function of ξ_l, can be calculated [75] using the tabulated optical data [36] and the Kramers-Kronig relation (19). In the original Ref. [15] it was assumed that the fused silica is a true dielectric of infinitely high resistivity. Thus, the dc conductivity of the plate was disregarded. Under this assumption the obtained theoretical results [15] are shown by the solid line in Fig. 7(a). It is seen that the Lifshitz theory of the Casimir-Polder interaction with the dc conductivity of fused silica disregarded is consistent with the measurement data.

It should be noted, however, that although fused silica is a good insulator, it possesses a nonzero conductivity at $T_p = 310\,$K. This conductivity is ionic in nature and is determined by the concentration of impurities (alkali ions). In Ref. [16] the computations of Ref. [15] were repeated both disregarding and including the dc conductivity of fused silica. When the dc conductivity was disregarded, the results of Ref. [15] given by the solid line were reproduced. When, however, the dc conductivity was included, quite different results were obtained shown by the dashed line in Fig. 7(a). It is seen that with inclusion of the dc conductivity the Lifshitz theory is inconsistent with at least two experimental points at shortest separations.

The second and third repetitions of the experiment of Ref. [15] were performed with the

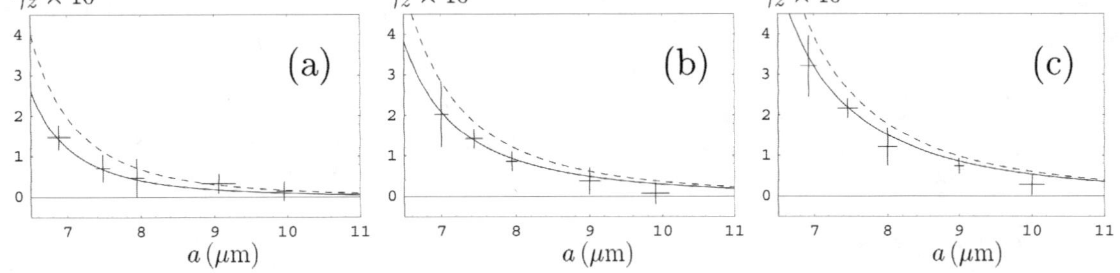

Figure 7. The fractional change in the trap frequency versus separation (a) in thermal equilibrium with $T_p = T_e = 310\,$K and out of thermal equilibrium (b) with $T_p = 479\,$K and $T_e = 310\,$K and (c) with $T_p = 605\,$K and $T_e = 310\,$K. Computations are done by neglecting (solid lines) and including (dashed lines) the dc conductivity of the dielectric plate. The experimental data are shown as crosses.

same temperature of environment, $T_e = 310\,\mathrm{K}$, but with increased temperatures of the plate, $T_p = 479\,\mathrm{K}$ and $605\,\mathrm{K}$, respectively. The obtained experimental data are shown as crosses in Fig. 7(b) for $T_p = 479\,\mathrm{K}$ and Fig. 7(c) for $T_p = 605\,\mathrm{K}$. The computational results of Ref. [15] disregarding the dc conductivity of fused silica are shown as the solid lines in Fig. 7(b,c). They were reproduced in Ref. [16] under an assumption that the dc conductivity of fused silica is omitted. As is seen in Fig. 7(b,c), the solid lines are in a good agreement with the experimental data. In Ref. [16] computations of the relative frequency shift γ_z for the second and third repetitions of the experiment with increased temperature of the plate were performed taking the dc conductivity of fused silica into account. The computational results shown by the dashed lines in Fig. 7(b,c) differ dramatically from the results obtained with the dc conductivity disregarded. As can be seen in Fig. 7(b), the three experimental points for $T_p = 479\,\mathrm{K}$ exclude the dashed line and the other two only touch it. As to the dashed line in Fig. 7(c), it is seen that all data points at $T_p = 605\,\mathrm{K}$ exclude the theoretical prediction computed taking the dc conductivity of fused silica into account. Thus, measurements of the thermal Casimir force between ^{57}Rb atoms and fused silica plate suggest that in the application of the Lifshitz theory to dielectric materials the dc conductivity of these materials should be omitted.

In fact, the results of all three landmark experiments performed by three different experimental groups are in agreement. According to these results, the relaxation properties of free charge carriers are irrelevant to the phenomenon of dispersion forces and should be disregarded. For metals characterized by a high density of charge carriers, they should be accounted for by means of the plasma model. For dielectrics charge carriers should be ignored. If this phenomenological prescription [6, 8, 38, 70] is used, the Lifshitz theory agrees well with all available experimental data.

4. Possibilities to control the magnitude and sign of the Casimir force

Applications of the Casimir force to nanotechnology require some operational means to modify the force magnitude at a fixed separation distance and even change the force sign with varying separation. This would permit to realize the actuation of devices with the Casimir force at a nanoscale and, when necessary, avoid such undesired effects as pull-in and stiction mentioned in Sec. 1. Below we briefly discuss different possibilities to control the Casimir force proposed in the literature some of which were already tested experimentally. Special attention is paid to magnetic materials which could significantly widen the role of Casimir forces in applied science. Here, we do not discuss the Casimir force between metamaterials because the application of the Lifshitz formula to such materials is questionable, and corresponding experimental data are not yet available. In spite of repeated speculations on the possibility of Casimir repulsion, it was shown [76] that the Casimir force between two plates made of metamaterials and separated by a vacuum gap is always attractive.

4.1. Different ways of modifying dielectric properties

To modify the magnitude of the Casimir force, the dielectric properties of interacting bodies should be changed over a wide frequency region. The most apparent possibility of decreasing the magnitude of the Casimir force, as compared with the case of two metallic test bodies, is to make one of them of a semiconductor or dielectric (or at least to cover them with sufficiently thick dielectric coating). As an example, in the experiment of Refs. [77, 78] the Casimir force acting between a Au coated sphere and Si plate at $a = 100\,\mathrm{nm}$ in vacuum was of about 30% less than between the same Au sphere and a Au plate. In the experiment of Ref. [79] the Casimir forces acting between a Au sphere and two Si plates with significantly different dopant concentrations were sequentially measured. The density of charge carriers in the first plate was $n_a \approx 1.2 \times 10^{16}\,\mathrm{cm}^{-3}$ and in the second plate $n_b \approx 3.2 \times 10^{20}\,\mathrm{cm}^{-3}$. It was found that by choosing plates with such different concentrations of charge carriers the Casimir force acting between a Au sphere and a Si plate can vary by 5% and 8% at separations $a = 60$ and 100 nm, respectively. In Ref. [80] the measured values of the Casimir force between Au sphere and Au plate were compared with respective measurements resulted from a Au sphere and a glass plate covered with a 190 nm layer of indium tin oxide which is a conductor at low frequencies (with respect to the Casimir force this material was first discussed in Ref. [39]). Within the separation region from 50 to 150 nm the Casimir force between a sphere and a plate covered with indium tin oxide was shown to be 40%–50% less than between the same sphere and a Au plate.

The use of test bodies made of dielectric material or coated with dielectric layers allows to decrease the magnitude of the Casimir force but does not allow to change this magnitude periodically. The latter was realized in the optical modulation experiment of Refs. [13, 14] considered in Sec. 3.2. In this experiment, the differences of the Casimir forces in the presence and in the absence of laser light on a Si plate are equal to 3% and 6.5% of the Casimir force in the absence of light at separations $a = 100$ and 150 nm, respectively. It is important that here the magnitude of the Casimir force varies periodically as long as the laser operated in the pulse mode is switched on.

Optical modulation of the Casimir force in vacuum makes it possible to periodically change the magnitude of the Casimir force but does not allow to replace attraction with repulsion. Considerable opportunities for nanotechnology would be opened up if pulsating Casimir plates were possible, moving back and forth under the action of the Casimir force without use of micromechanical springs. This can be achieved only by using both attractive and repulsive Casimir forces. In this respect, it should be noted that there is the case where the Casimir repulsion is well understood, i.e., the repulsion between the two plates described by the dielectric permittivities $\varepsilon^{(1,2)}(i\xi)$ separated with a liquid layer having the dielectric permittivity $\varepsilon^{(0)}(i\xi)$ such that [5, 6, 8]

$$\varepsilon^{(1)}(i\xi) < \varepsilon^{(0)}(i\xi) < \varepsilon^{(2)}(i\xi). \tag{27}$$

The Casimir repulsion in three-layer systems was observed in Refs. [81, 82, 83, 84] (see also Refs. [85, 86] concerning additional difficulties arising for measurements of the Casimir force in liquids).

Recently it was shown [87] that the illumination of one (Si) plate in three-layer systems Au – ethanol – Si, Si – ethanol – Si, and α-Al_2O_3 – ethanol – Si with laser pulses can change the Casimir attraction to the Casimir repulsion and vice versa. This allows to combine the advantages of the optical modulation with the Casimir repulsion in three-layer systems. As a result, one can obtain attraction and repulsion interchangeably at regular intervals. Calculations show [87] that in the system Au – ethanol – Si the Casimir force is repulsive at $a > 160$ nm. The illumination of the Si plate with laser light changes this repulsion for attraction. In the system Si – ethanol – Si, the Casimir force between the Si plates is attractive in the absence of laser light. However, when one of the two Si plates is illuminated, the attraction is replaced with repulsion at separations $a > 175$ nm [87]. In both these systems the magnitude of the repulsive pressure is several times less than the magnitude of the attractive pressure at the same separation. However, in the three-layer system α-Al_2O_3 – ethanol – Si, where the Si plate is illuminated with laser pulses, the light-induced Casimir repulsion is of the same order of magnitude as the Casimir attraction. Specifically, in this system in the presence of light the Casimir repulsion arises at $a > 70$ nm. Thus, at $a = 110$ nm the attractive Casimir pressure in the absence of light is equal to about –75 mPa, whereas the repulsive pressure in the presence of light is equal to 50 mPa.

The dielectric properties undergo changes during different phase transitions. It was shown [88] that the change in the Casimir free energy associated with the phase transition of a metal to the superconducting state is very small. Nevertheless, according to Refs. [89, 90], the magnitude of this change is comparable to the condensation energy of a superconducting film and causes a measurable increase in the value of the critical magnetic field. It was also proposed [91] to measure the change of the Casimir force in a superconducting cavity due to a small change of temperature.

Considerable change of the dielectric properties occurs in semiconductor materials that undergo a dielectric-metal phase transition with an increase in temperature. It was proposed [92] to measure the change of the Casimir force acting between an Au-coated sphere and a vanadium dioxide (VO_2) film deposited on a sapphire substrate, where the VO_2 undergoes a phase transition. At $T = 68\,^{\circ}$C thin VO_2 films undergo an abrupt transition from a dielectric-type semiconductor to a metallic phase. During this phase transition the resistivity of the film decreases by a factor 10^4 from 10 to $10^{-3}\,\Omega\,$cm. The increase in temperature necessary for the phase transition can be induced by laser light. Thus, a setup similar to the one described in Sec. 3.2 can be used. Preliminary theoretical results using the optical data for VO_2 films show [92, 93] that the difference Casimir force between a Au sphere and VO_2 film after and before the phase transition is sensitive to the model of conductivity properties used in computations. Thus, this experiment may help in the resolution of the problems discussed in Secs. 2 and 3.

Another type of dielectric-metal phase transitions leading to the change of dielectric properties is a transition from amorphous to crystalline state in AgInSbTe. In Ref. [94] the gradient of the Casimir force between a Au coated sphere of $20.2\,\mu$m diameter and $1\,\mu$m thick AgInSbTe films deposited onto Al-coated Si substrates was measured using an atomic force microscope in dynamic mode. Half of the films were amorphous and the other half were annealed to the metallic crystalline phase. It was shown that differences in the gradient of the Casimir force between a sphere and an amorphous film on the one hand and between a sphere and a crystalline film on the other hand achieve 20% at separations of about 100 nm. The obtained experimental results were compared with computations using the Lifshitz theory and significant deviations were found. Taking into account that according to Ref. [94] the role of surface roughness was negligible at separations above 70 nm, these deviations might be explained by the influence of localized charges in the amorphous phase and by the use of the Drude model in the extrapolation of the optical data in the crystalline phase where AgInSbTe exhitits metallic conductivity (see Secs. 2 and 3).

4.2. Prospects of using magnetic materials

The possibility to modify the magnitude of the Casimir force and even to change attraction for repulsion has long been discussed in the literature. Specifically, in Ref. [95] the Casimir pressure between two magnetodielectric plates was investigated in the approximation of frequency-independent ε and μ. For sufficiently large μ repulsive forces were found. It was commented in the literature, however, that for real magnetic materials μ is nearly equal to unity in the range of frequencies, which gives major contribution to the Casimir force [96]. As a result, the magnitude of μ always remains far away from the values needed to get the Casimir repulsion. In fact the largest magnitude of $\mu(i\xi)$ is achieved at $\xi = 0$. For diamagnets [97] $|\mu(0) - 1| \sim 10^{-5}$ always. Thus, these materials cannot be used for modifying the magnitude of the Casimir force. For paramagnets, with the single exception of ferromagnets, the deviation of $\mu(0)$ from unity is also negligibly small at any temperature [97].

The subset of paramagnetic materials called *ferromagnets* requires special attention when the aim is to modify the Casimir force. For such materials $\mu(0) \gg 1$ at temperatures lower than the so-called *Curie* temperature T_c [97]. The values of $\mu(i\xi)$ for ferromagnets, however, quickly decrease with the increase of frequency. Thus, for ferromagnetic metals $\mu(i\xi)$ drops toward unity at frequencies of order $10^5\,$Hz and for ferromagnetic dielectrics at frequencies of order $10^9\,$Hz (note that the first Matsubara frequency ξ_1 at $T = 300\,$K is of order $10^{14}\,$Hz). Because of this, in all computations using Eq. (1) at, say, $T > 0.3\,$K one can put $\mu(i\xi_l) = 1$ at all $l \geq 1$ and include ferromagnetic properties only in the zero-frequency term with $l = 0$ [98, 99].

Recently the formalism of functional determinants discussed in Sec. 2 was applied to calculate the Casimir force acting between an ideal metal cylinder far away from a magnetodielectric plate [100]. Calculations were performed at $T = 0$. It was argued that for a cylinder remote sifficiently far from the plate, it is justified to use the static values $\varepsilon(0)$ and $\mu(0)$ for the dielectric permittivity and static permeability of the plate material. Under these assumptions, both attractive and repulsive Casimir force was obtained depending on the values of $\varepsilon(0)$ and $\mu(0)$. It should be remarked, however, that in order for magnetic properties to influence the Casimir force, the characteristic frequency $\omega_c = c/(2a)$ must be less than 10^9 Hz. Hence it follows that cylinder-plate separation must be larger than 15 cm. At such large separations any discussion concerning the sign of the Casimir force seems problematic, because this force is equal to zero for all practical purposes and is considerably dominated by Newtonian gravity. As was shown in Ref. [101], the calculational results for the Casimir force between magnetodielectric materials with frequency-dependent μ depend crucially on whether the zero-temperature or thermal Lifshitz formula is used.

The investigation of thermal Casimir interaction between two magnetodielectric plates with account of real material properties was performed in Refs. [98, 99]. As is clear from the foregoing, only ferromagnets may show noticeable effect on the Casimir force. With respect to the Casimir effect, however, it is not reasonable to consider parallel plates made of so-called *hard* ferromagnetic materials which possess spontaneous magnetization. The point is that the magnetic interaction between such plates far exceeds any conceivable Casimir force. Ferromagnetic materials discussed below are what is referred to as *soft* ferromagnetic materials, which do not possess spontaneous magnetization. It is well known also [97] that the magnetic permeability of ferromagnets depends on the applied magnetic field (the so-called *hysteresis*). Keeping in mind that no external magnetic field is applied to the Casimir plates and that the mean value of the fluctuating magnetic field is equal to zero, one should consider what is often referred to as *initial* permeability, i.e., $\mu(\boldsymbol{H} = 0)$.

Now we present a few computational results illustrating the impact of ferromagnetic materials on the magnitude and sign of the Casimir force. Computations of the Casimir pressure were performed using Eq. (1) at room temperature with ferromagnetic properties included in the zero-frequency term $l = 0$. The dielectric properties of ferromagnetic and nonmagnetic metallic plates are described either by the Drude model (15) or by the plasma model (17) and the obtained results are compared. Keeping in mind that ferromagnetic properties contribute only through the zero-frequency term, it is reasonable to consider not too short separations where the relative contribution of this term is more pronounced. Below we consider separation distances from 0.5 to 6 μm, where the role of core electrons in the dielectric response can be neglected.

First we consider the Casimir pressure between two plates made of the ferromagnetic metal Co with $\mu_{\rm Co}^{(n)}(0) = 70$ [102]. The Drude parameters of Co are the following [103]: $\omega_{p,{\rm Co}}^{(n)} = 3.97$ eV and $\gamma_{\rm Co}^{(n)} = 0.036$ eV. The computational results are presented as a ratio to the zero-temperature Casimir pressure between two nonmagnetic plates made of ideal metal at zero temperature

$$P_0(a) = -\frac{\pi^2}{240}\frac{\hbar c}{a^4}. \tag{28}$$

In Fig. 8, the solid and dashed lines show the values of $P_{\rm Co}/P_0$ as functions of separation computed using the dielectric permittivity (a) of the Drude model and (b) of the plasma model with account of magnetic properties and with magnetic properties disregarded, respectively. Quantitatively, the role of magnetic properties can be characterized by the ratio

$$\eta_{P,{\rm Co}}(a,T) = \frac{P_{\rm Co}^{\rm solid}(a,T) - P_{\rm Co}^{\rm dashed}(a,T)}{P_{\rm Co}^{\rm dashed}(a,T)}. \tag{29}$$

With the increase in separation from 0.5 to 2 μm and then to 6 μm, $\eta_{P,{\rm Co}}^{D}$ computed using the Drude model varies from 12% to 44% and to 92%, respectively. Thus, when the Drude model

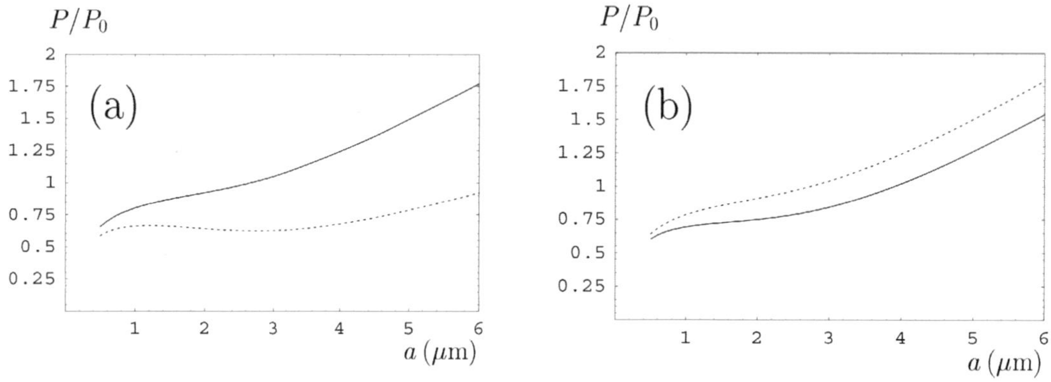

Figure 8. The relative Casimir pressures as functions of separation in the configuration of two parallel Co plates with account of magnetic properties (solid lines) and with magnetic properties disregarded (dashed lines). Computations are performed at $T = 300\,$K using (a) the Drude model and (b) the plasma model for the dielectric permittivity.

is used to describe the dielectric properties of a ferromagnetic metal, the magnetic properties markedly (up to almost two times at large separations) increase the magnitude of the Casimir pressure [98]. As can be seen in Fig. 8(b), for ferromagnetic metal described by the plasma model the impact of magnetic properties on the Casimir pressure is not so pronounced. Quantitatively, from Fig. 8(b) it follows that at separation 0.5, 2, and 6 μm $\eta_{P,\mathrm{Co}}^{p}$ varies from –6% to –17%, and to –14%, respectively. Mention should be made that if the plasma model is used, the inclusion of magnetic properties decreases the magnitude of the Casimir pressure. Similar results were also obtained for Fe [98]. In both cases the inclusion of the magnetic properties does not change the sign of the Casimir force leaving it attractive.

Very prospective materials for the investigation of the impact of magnetic properties on the Casimir force are ferromagnetic dielectrics. These materials, while displaying physical properties characteristic for dielectrics, demonstrate ferromagnetic behavior under the influence of an external magnetic field [104]. Many ferromagnetic dielectrics are, for instance, composite materials obtained on the basis of a matrix of a polymer compound with inclusion of nanoparticles of ferromagnetic metals. As was shown in Ref. [98], the Casimir force between two parallel plates made of ferromagnetic dielectric remains attractive, while the influence of magnetic properties on the Casimir pressure is much stronger than for Co plates considered above.

Here, we consider the configuration of one plate made of ferromagnetic dielectric ($n = 1$) and the other plate made of a nonmagnetic metal Au ($n = 2$). As is ahown below, in this configuration the results for the Casimir pressure are drastically different depending on whether the Drude or the plasma model is used for the description of Au. In the case when the Drude model is used, from Eqs. (4) and (15) it follows that $r_{\mathrm{TE}}^{(2)}(0, k_\perp) = 0$. Keeping in mind that at zero frequency magnetic properties do not influence the TM reflection coefficient, we arrive at the conclusion that if the Drude model is used for the description of Au there is no impact of magnetic properties on the Casimir pressure between gold and magnetodielectric plates. To perform computations when Au plate is described by the plasma model, one should choose some specific ferromagnetic dielecric. We consider the composite material on the basis of polystyrene with the volume fraction of ferromagnetic particles in the mixture $f = 0.25$. The magnetic permeability of such materials may vary over a wide range [105]. Below we use $\mu(0) = 25$. The dielectric permittivity of polystyrene $\varepsilon_d(i\xi)$ is presented in the form (10) with $K = 4$. The parameters of oscillators can be found in Ref. [34] leading to $\varepsilon_d(0) = 2.56$. As a result, the

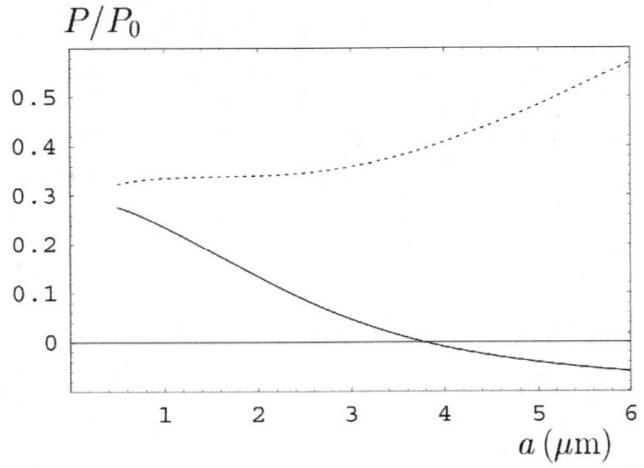

Figure 9. The relative Casimir pressures as functions of separation at $T = 300\,\mathrm{K}$ in the configuration of one plate made of ferromagnetic dielectric and the other plate made of Au with account of magnetic properties (solid line) and with magnetic properties disregarded (dashed line). Computations are performed using the plasma model for the dielectric permittivity of Au.

dielectric permittivity of the used ferromagnetic dielectric can be presented in the form [106]

$$\varepsilon_{\rm fd}^{(1)}(i\xi) = \varepsilon_d(i\xi)\left(1 + \frac{3f}{1-f}\right). \tag{30}$$

The dielectric permittivity of Au is described by Eq. (17) with $n = 2$, $\omega_p^{(2)} = 9.0\,\mathrm{eV}$.

The computational results for the relative Casimir pressure P/P_0 at $T = 300\,\mathrm{K}$ as a function of separation are presented in Fig. 9. The solid (dashed) lines show the results obtained with magnetic properties of ferromagnetic plate included (disregarded). As is seen in Fig. 9, there is the profound effect of magnetic properties of ferromagnetic dielectric on the Casimir pressure. Thus, at separations of 0.5, 2, and $6\,\mu$m the values of $\eta^p_{P,\rm fd-Au}$ are equal to –14%, –60%, and –110%, respectively. What is more important, the Casimir pressure P (and, thus, the Casimir force acting between parallel plates through a vacuum gap) changes sign and becomes positive at separations $a > 3.8\,\mu$m. This means that the Casimir force becomes repulsive if Au plate is described by the plasma model. The effect of repulsion between a nonmagnetic metal plate and a ferromagnetic dielectric plate can be used as an experimental test for the influence of magnetic properties on the Casimir force and for the model of dielectric permittivity of a metal plate.

As was mentioned in the beginning of this section, at the Curie temperature T_C specific for each material, ferromagnets undergo a phase transition [107]. At higher temperature they become paramagnets with negligibly small magnetic properties with respect to the Casimir force. The Curie temperature varies over a wide temperature region (for Co, for instance, it is equal to 1388 K). Keeping in mind to investigate the behavior of the Casimir pressure under the magnetic phase transition, it is preferable to deal with two plates made of ferromagnetic metal possessing the Curie temperature close to room temperature, such as Gd (T_C is of about 290 K). Dielectric properties of Gd are described by either the Drude or the plasma model with the parameters [108] $\omega_{p,\rm Gd} = 9.1\,\mathrm{eV}$ and $\gamma_{\rm Gd} = 0.58\,\mathrm{eV}$. The dependence of static magnetic permeability of Gd as a function of temperature in the vicinity of Curie temperature in modeled in Fig. 10(a) using the data of Ref. [109].

Computations of the relative Casimir pressure P/P_0 in the configuration of two Gd plates were performed at $a = 500\,\mathrm{nm}$ as a function of temperature using Eq. (1). The computational results are presented in Fig. 10(b), where the pairs of solid and dashed lines marked 1 and 2 indicate the results computed using the Drude and plasma models, respectively. As before, the solid lines take into account and the dashed lines disregard the magnetic properties of Gd. As can

Journal of Physics: Conference Series **258** (2010) 012001 doi:10.1088/1742-6596/258/1/012001

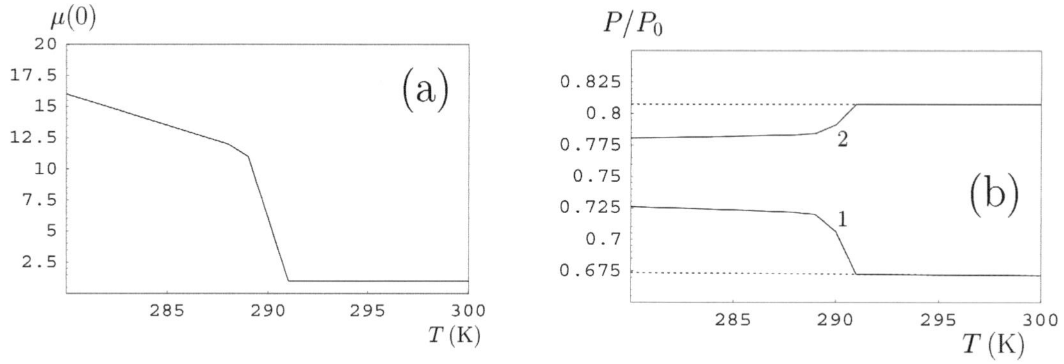

Figure 10. (a) The static magnetic permeability of Gd in the magnetic phase transition as a function of temperature. (b) The relative Casimir pressures as functions of temperature in the configuration of two parallel Gd plates at the separation $a = 0.5\,\mu$m. The solid and dashed lines take into account and disregard the magnetic properties, respectively. The pairs of lines marked 1 and 2 indicate respective computational results obtained using the Drude and plasma models.

be seen in Fig. 10(b), at $T > T_C$ the magnetic properties do not influence the Casimir pressure, whereas the Drude and plasma model approaches lead to results differing for about −19.5%. At $T < T_C$ the magnetic properties influence the Casimir pressure. Quantitatively, at $T = 280\,$K the relative influence of magnetic properties on the Casimir pressure is $\eta_{P,\mathrm{Gd}}^{D} = 7.4\%$ for the Drude model and $\eta_{P,\mathrm{Gd}}^{p} = -3.3\%$ for the plasma model. With account of magnetic properties, the relative difference between the predictions of the Drude and plasma model approaches for the Casimir pressure is approximately equal to −7%. Thus, the magnetic phase transition provides additional opportunities for the invistigation of the impact of magnetic properties on the Casimir force and for the selection between different models of dielectric permittivity.

To conclude this section, we note that possible influence of magnetic properties of ferromagnets on the Casimir force may be considered as somewhat analogous to the influence of real drift current of conduction electrons discussed in Sec. 2. If it is assumed that the fluctuating electromagnetic field can initiate such a current, we arrive at the Drude model approach to the thermal Casimir force which is considered as most natural by some authors [41, 42, 43]. This approach, however, was found to be in contradiction with the results of several experiments [9, 10, 11, 12, 13, 14, 15, 16]. In this connection the problem arises whether the fluctuating electromagnetic field can lead to magnetic effects in ferromagnets which remains to be solved.

5. Lateral Casimir force at the nanoscale

The most universally known *normal* Casimir force acts in the direction perpendicular to the interacting surfaces. However, when the material surfaces of the interacting bodies are anisotropic or they are positioned asymmetrically, a lateral Casimir force may exist, which acts tangential to the surface [17]. Experimentally the lateral Casimir force was first measured [18, 19] in the configuration of a sinusoidally corrugated Au-coated sphere above a sinusoidally corrugated Au-coated plate with equal corrugation periods. The measurement data were compared with theoretical predictions using the PFA. For two parallel ideal metal plates covered with uniaxial sinusoidal corrugations of equal periods the lateral Casimir force was calculated using the exact theory in the second perturbation order with respect to relative corrugation amplitude [110]. For sinusoidally corrugated plates made of real metals described by the plasma model, the lateral Casimir force was calculated in Refs. [111, 112] also in the second-order perturbation theory. Experimentally, the lateral Casimir force becomes measurable only for sufficiently large amplitudes of corrugations and at sufficiently short separations. Because of this, calculational methods restricted by the second-order perturbation theory are usually not sufficient for the comparison with the measurement data [113].

In the first measurement of the lateral Casimir force [18, 19] the corrugation amplitudes were $A_1 = 59\,\mathrm{nm}$ on the plate and $A_2 = 8\,\mathrm{nm}$ on the sphere. The period of corrugations was equal to $\Lambda = 1200\,\mathrm{nm}$, i.e., much larger than the separation between the sphere and the plate. The lateral Casimir force with an amplitude of $3.2 \times 10^{-13}\,\mathrm{N}$ at the shortest separation was found to sinusoidally oscillate as a function of the phase shift φ between the corrugations. It was also predicted theoretically, but not observed experimentally due to the use of insufficiently large corrugation amplitudes, that the lateral Casimir force is in fact asymmetric, i.e., deviates from purely sinusoidal harmonic dependence on the phase shift (this phenomenon is absent when computations are performed up to the second perturbation order).

In the recently performed new measurements of the lateral Casimir force between the Au-coated sphere and plate covered with uniaxial sinusoidal corrugations, the corrugation amplitudes were $A_1 = 85.4\,\mathrm{nm}$ on the plate and $A_2 = 13.7\,\mathrm{nm}$ on the sphere ($\tilde{A}_2 = 25.5\,\mathrm{nm}$ in the second set of measurements). The corrugations had an average period $\Lambda = 574.7\,\mathrm{nm}$, i.e., more than two times smaller than in Refs. [18, 19] in order to achieve the regime where the PFA becomes inapplicable to the calculation of the lateral Casimir force [see Fig. 11(a,b) where the corrugated surfaces of the plate and of the sphere used in the first set of measurements, respectively, are shown]. To obtain corrugations on the sphere with exactly the same period, as on the plate, a homogeneous sphere was sandwiched against the plate and pressed by means of a piezo. In such a manner sinusoidal corrugations from the plate were imprinted on the sphere. This was done preserving the sphericity of a sphere (see Fig. 12, where the lighter tone shows higher points on the surface of the sphere used in the first set of measurements).

Taking into account the sphericity of the second body and sinusoidal corrugations on both bodies in the framework of the PFA and using the Lifshitz formula (1), one arrives to the following approximate expression for the lateral Casimir force [21]

$$F_{\mathrm{lat}}(a,T,\varphi) = \frac{\pi k_B T R A_1 A_2}{2a^3\Lambda\beta}\sin\varphi \sum_{n=1}^{\infty}\int_{\zeta_l}^{\infty} y\,dy\,e^{-ny}\,I_1(n\beta y)\left[r_{\mathrm{TM}}^{2n}(i\zeta_l,y)+r_{\mathrm{TE}}^{2n}(i\zeta_l,y)\right]. \quad (31)$$

Here,

$$\beta \equiv \beta(a,\varphi) = \frac{1}{a}(A_1^2 + A_2^2 - 2A_1A_2\cos\varphi)^{1/2}, \quad (32)$$

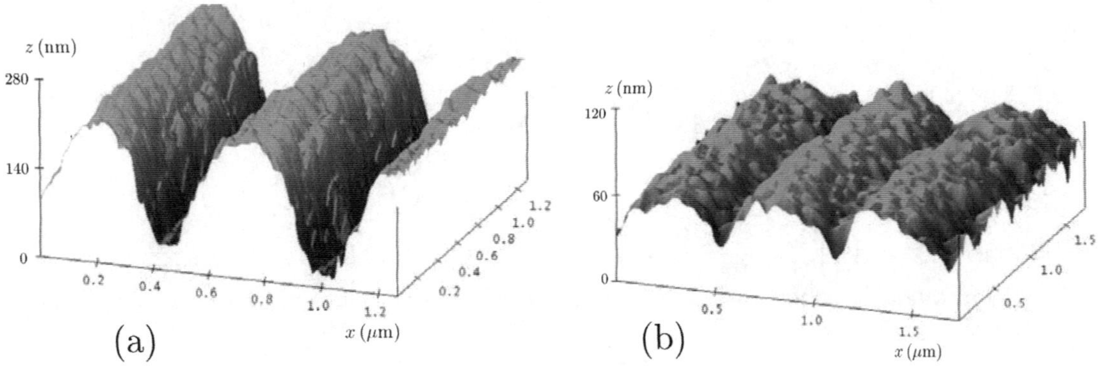

Figure 11. An AFM scan (a) of the plate surface and (b) of the surface of the sphere showing the sinusoidal corrugations covered with stochastic roughness.

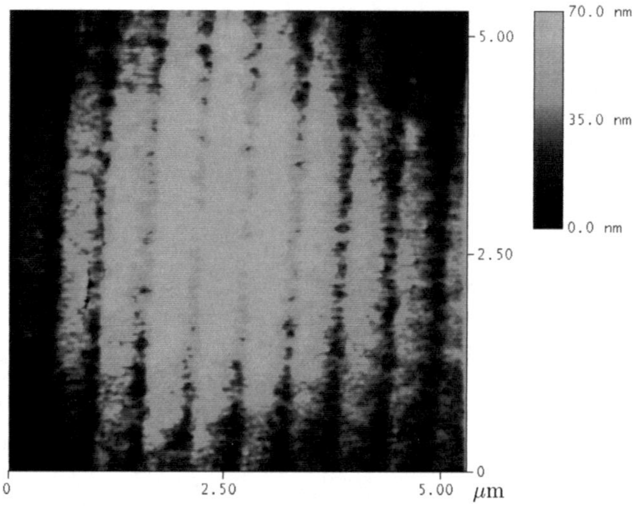

Figure 12. The imprinted corrugations on the sphere. The lighter area shows higher points and hence demonstrates the sphericity of the imprinted surface.

$I_1(z)$ is the Bessel function of an imaginary argument, and the following dimensionless variables are used:

$$y = 2aq_l, \qquad \zeta_l = \frac{2a\xi_l}{c}. \tag{33}$$

Remembering that Eq. (31) uses the PFA for the description of sinusoidal corrugations, it should lead to sufficiently exact results under a condition $2\pi a \ll \Lambda$ [6, 8]. As to the applicability of the PFA to sphere-plate geometry, this is garanteed with large safety margin because $a \ll R = 97\,\mu$m.

Measurements of the lateral Casimir force in Refs. [20, 21] were performed at separations between the mean levels of corrugations of about 100–200 nm. Because of this, the application condition of the PFA for the description of corrugations was not satisfied. Therefore, some corrections to Eq. (31) due to diffraction-type effects were expected to arise. To take the diffraction-type effects into account, the exact description of corrugations using the Rayleigh scattering theory was developed [20, 21] while preserving the use of the PFA for the description of sphere-plate geometry, where it is sufficiently accurate. As a result, the following "exact" expression for the lateral Casimir force was obtained [20, 21]

$$\begin{aligned} F_{\rm lat}^{\rm exact}(a,T,\varphi) &= \frac{4k_BTR}{\Lambda}\int_a^\infty dz {\sum_{l=0}^{\infty}}{}' \frac{\partial}{\partial\varphi}\int_0^\infty dk_y \int_0^{\pi/\Lambda} dk_x \\ &\times \ln\det\left[I - R_{\rm bot}^{(1)}(\boldsymbol{k}_\perp, i\xi_l)R_{\rm top}^{(2)}(\boldsymbol{k}_\perp, i\xi_l, z+A_1+A_2,\varphi)\right]. \end{aligned} \tag{34}$$

Here, $R_{\rm bot}^{(1)}$ ($R_{\rm top}^{(2)}$) is the reflection matrix of the downward (upward) moving waves on the bottom (top) grating under a condition that the top (bottom) grating is removed, and I is an infinite dimensional unit matrix. In both Eqs. (31) and (34) the dielectric properties of Au were taken into account by means of the generalized plasma-like model (24).

In Fig. 13(a) we show the measured data in the first set of measurements (dots) and computed using the exact theory of Eq. (34) lateral Casimir force (solid line) over four corrugation periods versus the normalized lateral displacement x/Λ between the corrugated surfaces at a separation

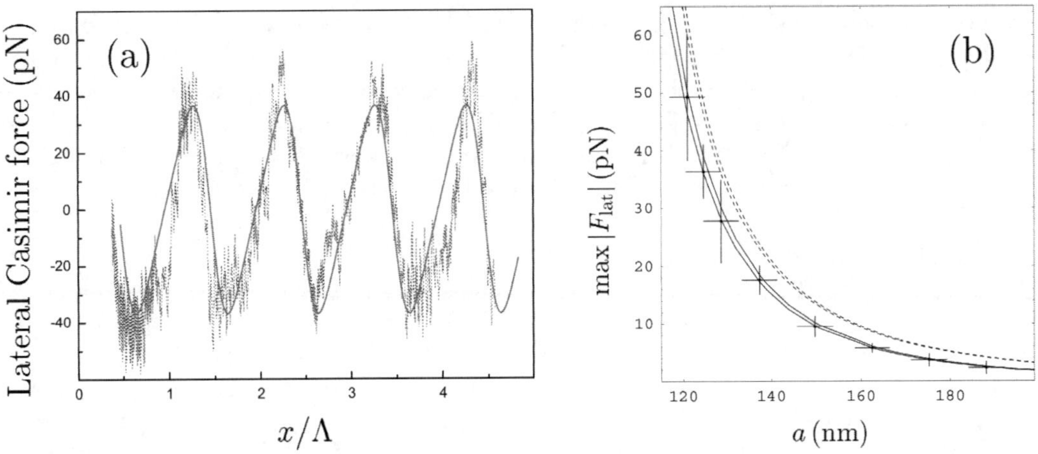

Figure 13. (a) The experiment (dots) and the exact theory (solid line) for the lateral Casimir force versus the lateral displacement normalized for the corrugation period at the separation 124.7 nm. No fitting parameters are used. (b) The experimental data (crosses) and theoretical values computed using the exact theory (the band between the solid lines) and using the proximity force approximation (the band between the dashed lines) for the maximum magnitudes of the lateral Casimir force versus separation. No fitting parameters are used.

$a = 124.7$ nm (similar figures related to larger separations and to the second set of measurements can be found in Ref. [21]). As can be seen in Fig. 13(a), the lateral Casimir force is a periodic function of the phase shift. The important characteristic feature of the periodic curve shown in the figure is that it is *asymmetric*, i.e., the dependence of $F_{\rm lat}$ on x/Λ is not strictly sinusoidal (note that the sinusoidal dependence of the lateral force on $\varphi = 2\pi x/\Lambda$ holds only if the calculations are restricted to the second order in corrugation amplitudes $\sim A_1A_2$). The true asymmetry of the measured lateral force is obvious even without the theory line. For example, in Fig. 13(a) the average shift of each maximum point from the midpoint of two adjacent minima is $(0.12 \pm 0.01)\Lambda$. We emphasize that the theoretical solid line was obtained with no fitting parameters. As can be seen in Fig. 13(a), the exact theory of Eq. (34) is in a very good agreement with the experimental data shown as dots.

The experimental data for the max $|F_{\rm lat}|$ versus separation a are shown in Fig. 13(b) as crosses. The arms of crosses indicate the total experimental error determined at a 95% confidence level for each experimental point individually. The exact computational results for the max $|F_{\rm lat}|$ are shown in Fig. 13(b) as a band between the two solid lines versus separation. The width of this band takes into account the computational errors and the correction to Eq. (34) due to surface roughness. As is seen in Fig. 13(b), the exact theory is in a very good agreement with the measurement data. The computational results for the max $|F_{\rm lat}|$ at different separations obtained using the PFA [Eq. (31)] are shown in Fig. 13(b) as a band between two dashed lines. The width of this band takes into account the role of surface roughness. As can be seen in Fig. 13(b), the experimental data indicated as crosses are inconsistent with the prediction of the PFA approach, thus revealing the role of diffraction-type effects for corrugations of relatively small period used in this experiment.

Besides fundamental interest in the phenomenon of lateral Casimir force as one more manifestation of the zero-point oscillations of electromagnetic field, it is pertinent to note that this version of the Casimir force gives the possibility to actuate lateral translations of corrugated surfaces in micromachines. The lateral Casimir force might be used to solve the tribological problems plaguing the microdevice industry. Specifically, it was proposed to use this effect for the frictionless transmission of lateral motion by means of a nanoscale rack and pinion without intermeshing cogs [114, 115, 116] or in a ratchet with asymmetric corrugations driven by the Casimir force [117]. Such devices could be used for transfering motion between two microelectromechanical systems without mechanical contact. The experimental and theoretical results on the lateral Casimir force discussed above bring us closer to the realization of micromachines driven by the zero-point fluctuations of quantum vacuum.

6. Conclusions and discussion

In the foregoing, we have listed several experiments and theoretical advances in the Casimir effect which has come to the center of public attention very recently. The Casimir effect deals with quantum vacuum, which is the most fundamental and puzzling type of physical reality. This is a possible reason why the research area in this field is so wide. On the one hand, it is closely connected with foundations of quantum statistical physics and even calls for reconsideration of some of its basic concepts. On the other hand, the Casimir effect quickly turns into a practical tool of nanotechnology which is expected to find numerous applications in the next generation of microdevices.

As was discussed in Sec. 2, the fundamental Lifshitz theory of the van der Waals and Casimir forces comes into conflict with basics of thermodynamics when the relaxation properties of free charge carriers are taken into account. In Sec. 3, the three landmark recently performed experiments are described demonstrating that in a literal sense, the Lifshitz theory is in a deep contradiction with the measurement data. Only if we accept the phenomenological prescription that for dielectrics charge carriers are disregarded and for metals they are described by the plasma model, the theoretical predictions of the Lifshitz theory come in agreement with the data. This situation cannot be considered as satisfactory. In fact we deal with some kind of a paradox. The mathematical formalism of the Lifshitz theory during the last few years has undergone far reaching generalization discussed in Secs. 2 and 5. It is presently possible to write the Lifshitz-type formulas for compact objects of arbitrary shape in terms of reflection amplitudes of electromagnetic oscillations on their surfaces. Based on this it was even believed that at the present time computation of the Casimir force between any bodies is only a problem of sufficient computer facilities. As was argued in Secs. 2 and 3, the problem is, however, much deeper. In fact the reflection amplitudes suggested by our knowledge of physical processes with real (not fluctuating) electromagnetic fields lead to computational results in contradiction with the experimental data making powerful mathematical formalism at least insufficient. This is a fundamental problem to be solved in the future.

From the point of view of applications in nanotechnology, large progress on the control the magnitude of the Casimir force was achieved. This was done by means of a modification of dielectric properties by using different materials and different influential factors such as irradiation with laser light. In several experiments considered in Sec. 4.1 it was possible to modify the magnitude of the Casimir force from a few percent to about 50%. There were several proposals on how to achieve the Casimir repulsion through the vacuum gap, but to the present day the Casimir repulsion was realized only in the familiar three-layer system. One more opportunity to control the magnitude of the Casimir force is suggested by phase transitions. In Sec. 4.1 we have discussed the phase transition from dielectric to the metallic state. A few tens of percent variation in the magnitude of the Casimir force is achievable under a phase transition of the plate material.

Quick progress in the application of the Casimir force in nanotechnology might be caused by the use of ferromagnetic materials. For real material properties this subject was investigated only recently (see Sec. 4.2). What is most striking, according to the Lifshitz theory the Casimir interaction of a nonmagnetic metallic plate described by the plasma model with a ferromagnetic dielectric plate through the vacuum gap becomes repulsive under some conditions. This effect, if exists in nature, opens up the way for the developments of various microdevices. We should notice, however, that the influence of magnetic properties on the Casimir interaction is not yet confirmed experimentally. The confirmation or exclusion of such an influence is expected in the near future.

One more opportunity for applications in nanotechnology is suggested by the lateral Casimir force discussed in Sec. 5. Together with the normal Casimir force, it gives the possibility to actuate both normal and lateral translations of corrugated surfaces in micromachines by means of the electromagnetic zero-point oscillations. With respect to tribological problems, the lateral Casimir force allows frictionless transition of lateral motion. All the above allows to conclude that the Casimir effect promises important breakthroughs in both fundamental physics and its applications in nanotechnology.

Acknowledgments

The author is grateful to the Organizing Committee of ICSF2010 for the invitation to present this talk, kind hospitality during the Conference and financial support. CNPq (Brazil), process 310039/2010–0 is also acknowledged for partial financial support. I am grateful to all co-authors of joint papers for helpful discussions and collaboration.

[1] Casimir H B G 1948 *Proc. K. Ned. Akad. Wet.* **51** 793
[2] Buks E and Roukes M L 2001 *Phys. Rev.* B **63** 033402
[3] Chan H B, Aksyuk V A, Kleiman R N, Bishop D J and Capasso F 2001 *Science* **291** 1941
[4] Lifshitz E M 1956 *Zh. Eksp. Teor. Fiz.* **29** 94 (*Sov. Phys. JETP* **2** 73)
[5] Dzyaloshinskii I E, Lifshitz E M and Pitaevskii L P 1961 *Sov. Phys. Usp.* **4** 153
[6] Klimchitskaya G L, Mohideen U and Mostepanenko V M 2009 *Rev. Mod. Phys.* **81** 1827
[7] Mostepanenko V M and Klimchitskaya G L 2010 *Int. J. Mod. Phys.* A **25** 2302
[8] Bordag M, Klimchitskaya G L, Mohideen U and Mostepanenko V M 2009 *Advances in the Casimir Effect* (Oxford: Oxford University Press)
[9] Decca R S, Fischbach E, Klimchitskaya G L, Krause D E, López D and Mostepanenko V M 2003 *Phys. Rev.* D **68** 116003
[10] Decca R S, López D, Fischbach E, Klimchitskaya G L, Krause D E and Mostepanenko V M 2005 *Ann. Phys. NY* **318** 37
[11] Decca R S, López D, Fischbach E, Klimchitskaya G L, Krause D E and Mostepanenko V M 2007 *Phys. Rev.* D **75** 077101
[12] Decca R S, López D, Fischbach E, Klimchitskaya G L, Krause D E and Mostepanenko V M 2007 *Eur. Phys. J.* C **51** 963
[13] Chen F, Klimchitskaya G L, Mostepanenko V M and Mohideen U 2007 *Optics Express* **15** 4823
[14] Chen F, Klimchitskaya G L, Mostepanenko V M and Mohideen U 2007 *Phys. Rev.* B **76** 035338
[15] Obrecht J M, Wild R J, Antezza M, Pitaevskii L P, Stringari S and Cornell E A 2007 *Phys. Rev. Lett.* **98** 063201
[16] Klimchitskaya G L and Mostepanenko V M 2008 *J. Phys. A: Math. Theor.* **41** 312002
[17] Golestanian R and Kardar M 1997 *Phys. Rev. Lett.* **78** 3421
[18] Chen F, Mohideen U, Klimchitskaya G L and Mostepanenko V M 2002 *Phys. Rev. Lett.* **88** 101801
[19] Chen F, Mohideen U, Klimchitskaya G L and Mostepanenko V M 2002 *Phys. Rev.* A **66** 032113
[20] Chiu H-C, Klimchitskaya G L, Marachevsky V N, Mostepanenko V M and Mohideen U 2009 *Phys. Rev.* B **80** 121402(R)
[21] Chiu H-C, Klimchitskaya G L, Marachevsky V N, Mostepanenko V M and Mohideen U 2010 *Phys. Rev.* B **81**, 115417
[22] Bimonte G, Klimchitskaya G L and Mostepanenko V M 2009 *Phys. Rev.* A **79** 042906
[23] Zhou F and Spruch L 1995 *Phys. Rev.* A **52** 297
[24] Tomaš M S 2002 *Phys. Rev.* A **66** 052103
[25] Raabe C, Knöll L and Welsch D-G 2003 *Phys. Rev.* A **68** 033810
[26] Emig T, Jaffe R L, Kardar M and Scardicchio A 2006 *Phys. Rev. Lett.* **96** 080403
[27] Bulgac A, Magierski P and Wirzba A 2006 *Phys. Rev.* D **73** 025007
[28] Kenneth O and Klich I 2006 *Phys. Rev. Lett.* **97** 160401
[29] Bordag M 2006 *Phys. Rev.* D **73** 125018
[30] Kenneth O and Klich I 2008 *Phys. Rev.* B **78** 014103
[31] Bordag M and Pirozhenko I 2010 *Phys. Rev.* D **81** 085023

[32] Wilson A G 1931 *Proc. Roy. Soc. Lond.* A **133** 458
[33] Mott N F 1990 *Metal-Insulator Transitions* (London: Taylor and Francis)
[34] Parsegian V A 2005 *Van der Waals forces: A Handbook for Biologists, Chemists, Engineers, and Physicists* (Cambridge: Cambridge University Press)
[35] Inui N 2003 J. Phys. Soc. Jap. **72** 2198
[36] Palik E D (ed) 1985 *Handbook of Optical Constants of Solids* (New York: Academic)
[37] Geyer B, Klimchitskaya G L and Mostepanenko V M 2005 *Phys. Rev.* D **72** 085009
[38] Klimchitskaya G L and Geyer B 2008 *J. Phys. A.: Mat. Theor.* **41** 164032
[39] Bezerra V B, Klimchitskaya G L and Mostepanenko V M 2002 *Phys. Rev.* A **66** 062112
[40] Bezerra V B, Klimchitskaya G L, Mostepanenko V M and Romero C 2004 *Phys. Rev.* A **69** 022119
[41] Boström M and Sernelius B E 2004 *Physica* A **339** 53
[42] Høye J S, Brevik I, Ellingsen S A and Aarseth J B 2007 *Phys. Rev.* E **75** 051127
[43] Brevik I, Ellingsen S A, Høye J S and Milton K A 2008 *J. Phys. A: Math. Theor.* **41** 164017
[44] Klimchitskaya G L and Mostepanenko V M 2008 *Phys. Rev.* E **77** 023101
[45] Klimchitskaya G L, Mohideen U and Mostepanenko V M 2008 *J. Phys. A: Math. Theor.* **41** 432001
[46] Pitaevskii L P 2008 *Phys. Rev. Lett.* **101** 163202
[47] Dalvit D A R and Lamoreaux S K 2008 *Phys. Rev. Lett.* **101** 163203
[48] Chazalviel J-N 1999 *Coulomb Screening of Mobile Charges: Applications to Material Science, Chemistry and Biology* (Boston: Birkhauser)
[49] Geyer B, Klimchitskaya G L, Mohideen U and Mostepanenko V M 2009 *Phys. Rev. Lett.* **102** 189301
[50] Mostepanenko V M, Decca R S, Fischbach E, Geyer B, Klimchitskaya G L, Krause D E, López D and Mohideen U 2009 *Int. J. Mod. Phys.* A **24** 1721
[51] Klimchitskaya G L 2009 *J. Phys. Conf. Ser.* **161** 012002
[52] Dure J C, Maass P, Roling B and Sidebottom D L 2009 *Rep. Progr. Phys.* **72** 046501
[53] Shklovskii B I and Efros A L 1984 *Electronic Properties of Doped Semiconductors. Solid State Series*, v.45 (Berlin: Springer)
[54] Decca R S, Fischbach E, Geyer B, Klimchitskaya G L, Krause D E, López D, Mohideen U and Mostepanenko V M 2009 *Phys. Rev. Lett.* **102** 189303
[55] Mostepanenko V M, 2009 *J. Phys. Conf. Ser.* **161** 012003
[56] Pitaevskii L P 2009 *Phys. Rev. Lett.* **102** 163203
[57] Dalvit D A R and Lamoreaux S K 2009 *Phys. Rev. Lett.* **102** 163204
[58] Krause D E, Decca R S, López D and Fischbach E 2007 *Phys. Rev. Lett.* **98** 050403
[59] Canaguier-Durand A, Maia Neto P A, Lambrecht A and Reynaud S 2010 *Phys. Rev. Lett.* **104** 040403
[60] Canaguier-Durand A, Maia Neto P A, Lambrecht A and Reynaud S 2010 *Phys. Rev.* A **82** 012511
[61] Zandi R, Emig T and Mohideen U 2010 *Phys. Rev.* B **81** 195423
[62] Lambrecht A and Reynaud S 2000 *Eur. Phys. J.* D **8** 309
[63] Pirozhenko I, Lambrecht A and Svetovoy V B 2006 *New J. Phys.* **8** 238
[64] Svetovoy V B, van Zwol P J, Palasantzas G and Hosson J Th M 2008 *Phys. Rev.* B **77** 035439
[65] Decca R S, López D and Osquiguil E 2010 *Int. J. Mod. Phys.* A **25** 2223
[66] Geyer B, Klimchitskaya G L and Mostepanenko V M 2010 *Phys. Rev.* B **81** 245421
[67] Inui N 2004 J. Phys. Soc. Jap. **73** 332
[68] Inui N 2006 J. Phys. Soc. Jap. **75** 024004
[69] Inui N 2007 J. Phys. Conf. Ser. **89** 012018
[70] Mostepanenko V M and Geyer B 2008 *J. Phys. A: Mat. Theor.* **41** 164014
[71] Shapiro B 2010 *Phys. Rev.* B **82** 075205
[72] Antezza M, Pitaevskii L P and Stringari S 2004 *Phys. Rev.* A **70** 053619
[73] Antezza M, Pitaevskii L P and Stringari S 2005 *Phys. Rev. Lett.* **95** 113202
[74] Harber D M, Obrecht J M, McGuirk J M and Cornell E A 2005 *Phys. Rev.* A **72** 033610
[75] Caride A O, Klimchitskaya G L, Mostepanenko V M and Zanette S I 2005 *Phys. Rev.* A **71** 042901
[76] Silveirinha M G 2010 *Phys. Rev.* B **82** 085101
[77] Chen F, Mohideen U, Klimchitskaya G L and Mostepanenko V M 2005 *Phys. Rev.* A **72** 020101(R)
[78] Chen F, Mohideen U, Klimchitskaya G L and Mostepanenko V M 2006 *Phys. Rev.* A **74** 022103
[79] Chen F, Klimchitskaya G L, Mostepanenko V M and Mohideen U 2006 *Phys. Rev. Lett.* **97** 170402
[80] deMan S, Heeck K, Wijngaarden R J and Iannuzzi D 2009 *Phys. Rev. Lett.* **103** 040402
[81] Milling A, Mulvaney P and Larson I 1996 *J. Colloid Interface Sci.* **180** 460
[82] Meurk A, Luckham P F and Bergström L 1997 *Langmuir* **13** 3896
[83] Lee S-W and Sigmund M W 2002 *Colloids Surf.* A **204** 43
[84] Munday J N, Capasso F and Parsegian V A 2009 *Nature* **457** 170
[85] Geyer B, Klimchitskaya G L, Mohideen U and Mostepanenko V M 2008 *Phys. Rev.* A **77** 036102

[86] van Zwol P J, Palasantzas G and Hosson J Th M 2009 *Phys. Rev.* B **79** 195428
[87] Klimchitskaya G L, Mohideen U and Mostepanenko V M 2007 *J. Phys. A.: Mat. Theor.* **40** F841
[88] Mostepanenko V M and Trunov N N 1997 *The Casimir Effect and Its Applications* (Oxford: Oxford University Press)
[89] Bimonte G, Calloni E, Exposito G, Milano L and Rosa L 2005 *Phys. Rev. Lett.* **94** 180402
[90] Bimonte G, Calloni E, Exposito G and Rosa L 2005 *Nucl. Phys.* B **726** 441
[91] Bimonte G 2008 *Phys. Rev.* A **78** 062101
[92] Castillo-Garza R, Chang C-C, Jimenez D, Klimchitskaya G L, Mostepanenko V M and Mohideen U 2007 *Phys. Rev.* A **75** 062114
[93] Pirozhenko I and Lambrecht A 2008 *Phys. Rev.* A **77** 013811
[94] Torricelli G, van Zwol P J, Shpak O, Binus C, Palasantzas G, Kooi B J, Svetovoy V B and Wuttig M 2010 *Phys. Rev.* A **82** 010101(R)
[95] Kenneth O, Klich I, Mann A and Revzen M 2002 *Phys. Rev. Lett.* **89** 033001
[96] Iannuzzi D and Capasso F 2003 *Phys. Rev. Lett.* **91** 029101
[97] Vonsovskii S V 1974 *Magnetism* (New York: J. Wiley).
[98] Geyer B, Klimchitskaya G L and Mostepanenko V M 2010 *Phys. Rev.* B **81** 104101
[99] Klimchitskaya G L, Geyer B and Mostepanenko V M 2010 *Int. J. Mod. Phys.* A **25** 229360
[100] Rahi S J, Emig T, Graham N, Jaffe R L and Kardar M 2009 *Phys. Rev.* D **80** 085021
[101] Tomaš M S 2005 *Phys. Lett.* A **342** 381
[102] Goldman A 1999 *Handbook of Modern Ferromagnetic Materials* (New York: Springer)
[103] Ordal M A, Bell R J, Alexander Jr R W, Long LL and Querry M R 1985 *Appl. Opt.* **24** 4493
[104] Chambers S A 2006 *Surf. Sci. Rep.* **61** 345
[105] Roy K and Datta A N 1974 *J. Phys.* D **7** 1053
[106] Lewin L 1947 *J. Inst. Electr. Eng.* Part 3 **94** 64
[107] Ausloos M (ed) 1983 *Magnetic Phase Transitions* (New York: Springer)
[108] Palik E D (ed) 1998 *Handbook of Optical Constants of Solids III* (New York: Academic)
[109] Coey J M D, Skumryev V and Gallagher K 1999 *Nature* **401** 35
[110] Emig T, Hanke A, Golestanian R and Kardar M 2003 *Phys. Rev.* A **67** 022114
[111] Rodrigues R B, Maia Neto P A, Lambrecht A and Reynaud S 2006 *Phys. Rev. Lett.* **96** 100402
[112] Rodrigues R B, Maia Neto P A, Lambrecht A and Reynaud S 2006 *Phys. Rev.* A **75** 062108
[113] Chen F, Mohideen U, Klimchitskaya G L and Mostepanenko V M 2006 *Phys. Rev. Lett.* **98** 068901
[114] Ashourvan A, Miri M and Golestanian R 2007 *Phys. Rev. Lett.* **98** 140801
[115] Ashourvan A, Miri M and Golestanian R 2007 *Phys. Rev.* E **75** 040103(R)
[116] Ashourvan A, Miri M and Golestanian R 2007 *J. Phys. Conf. Ser.* **89** 012017
[117] Emig T 2007 *Phys. Rev. Lett.* **98** 160801

Journal of Physics: Conference Series **258** (2010) 012002 doi:10.1088/1742-6596/258/1/012002

Numerical calculation of the Casimir forces between a gold sphere and a nanocomposite sheet

Norio Inui
Graduate School of Engineering, University of Hyogo, Shosha 2167, Himeji, Hyogo, 671-2201, Japan

E-mail: inui@eng.u-hyogo.ac.jp

Kouji Miura
Department of Physics, Aichi University of Education, Hirosawa 1, Igayacho, Kariya-shi, Aichi 448-8542, Japan

E-mail: kmiura@auecc.aichi-edu.ac.jp

Kensuke Akamatsu
Department of Nanobiochemistry, Konan University, 7-1-20 Minatojimaminami, Chuo-ku, Kobe 650-0047, Japan

E-mail: akamatsu@center.konan-u.ac.jp

Makoto Ishikawa
Department of Physics, Aichi University of Education, Hirosawa 1, Igayacho, Kariya-shi, Aichi 448-8542, Japan

E-mail: makoishi@auecc.aichi-edu.ac.jp

Abstract. The repulsive Casimir force is expected as a force which enables to levitate small objects such as machine parts used in Micro Electro Mechanical Systems (MEMS), and superlubricity in MEMS may be realized by this levitation. We study the Casimir force between a gold sphere and a nanocomposite sheet containing many nickel nanoparticles. In particular, we focus on the dependence of the Casimir force on the separation between the gold sphere and the surface of the nanocomposite sheet. The Casimir force changes from the attractive force to the repulsive force as the separation increases. The strength of the repulsive force is, however, too small to levitate MEMS parts.

1. Introduction

Realization of superlubricity is one of important targets in science of friction. Recently, quantum levitation by using the repulsive Casimir forces[1] has been proposed[2, 3]. The repulsive Casimir force is already observed in liquid[2, 3]. However, it is not observed in atmosphere, yet. Therefore, we need to develop the method of generating the strong repulsive Casimir in atmosphere in order to realize the quantum levitation in atmosphere.

Kenneth and Klich have proved that the Casimir interaction between two (non-magnetic) dielectric bodies related by reflection is always attractive, independently of the exact form of the bodies or their dielectric properties[4, 5]. Therefore, we must use magnetic-dielectric bodies[6] or break the reflection symmetry[7]. Boyer calculated the Casimir force between a perfectly conductive plate and a magnetic plate of which the magnetic permeable is infinitely and showed that the Casimir force between plates is repulsive[8]. Although materials with an infinite magnetic permeability does not exist, we expect that the repulsive Casimir force can be generated by increasing the magnetic permeability. Almost all natural materials exhibit no magnetic response for an electric-magnetic field in the range of visual light. Thus we need to investigate new materials, namely metamaterials[9, 10, 11, 12]

In this paper, we consider nanocomposite as a candidate of the magnetic body. Nanocomposite which is consisted of nickel nanoparticles has the magnetic permeability greater than one for nonzero frequency as a result of superparamagnetism[13]. If the size of nickel nanoparticle is much less than typical size of magnetic domain, the magnetization is uniform throughout the particle; accordingly the magnetization of a nanoparticle can be regarded as a single giant magnetic moment. The magnetization in a small nanoparticle can randomly flip direction under the influence of temperature. This results in the drastic reduction of the relaxation time, and nickel nanocomposite exhibits superparamagnetism in the infrared regime.

Nanocomposite containing nickel nanoparticle has already succeeded to be synthesized[14]. The important values characterizing superparamagnetism of the nanocomposite are the size of nanoparticle and the volume fraction. These parameters can be controlled independently by using polymers as a material of the matrix (the details of nanocomposite is in ref. [14]) We examine the dependence of the Casimir force on the separation between a gold sphere and a surface of the nanocomposite sheet containing nickel nanoparticles. Furthermore, we consider the dependence of the Casimir force on the volume fraction of nickel particle.

2. Casimir energy between a nickel plate and a gold plate

2.1. The Casimir energy between dielectric-magnetic plates

We begin to consider the Casimir energy between the infinite dielectric-magnetic plates which are separated by a gap a as shown in figure 1. According to the Lifshitz theory[15, 16, 1] the Casimir force between plates using the reflection coefficient $r_{\rm TM}$ and $r_{\rm TE}$ defined below as

$$E(a) = \frac{\hbar}{4\pi^2}\int_0^\infty k_\perp dk_\perp \int_0^\infty d\xi \left\{\ln\left[1 - r_{\rm TM}^{(1)} r_{\rm TM}^{(2)} e^{-2qa}\right] + \ln\left[1 - r_{\rm TE}^{(1)} r_{\rm TE}^{(2)} e^{-2qa}\right]\right\}, \qquad (1)$$

where $k_\perp$ is the magnitude of the wave vector parallel to the surface, ξ is the frequency variable along the imaginary axis ($\omega = i\xi$), and q is defined as

$$q(i\xi, k_\perp) = \sqrt{k_\perp^2 + \frac{\xi^2}{c^2}}. \qquad (2)$$

If the thickness of the metamaterial slab is infinite, its reflection coefficients are given by

$$\begin{aligned} r_{\rm TM}^{(i)} &= \frac{\epsilon_i(\omega)q - K_i}{\epsilon_i(\omega)q + K_i}, \\ r_{\rm TE}^{(i)} &= \frac{\mu_i(\omega)q - K_i}{\mu_i(\omega)q + K_i}, \end{aligned} \qquad (3)$$

where $\epsilon_i(\omega)$ and $\mu_i(\omega)$ are the dielectric function and the magnetic permeability of the metamaterial slab. Here, the function K_i for $i = 1, 2$ is defined as

$$K_i = \sqrt{k_\perp^2 + \epsilon_i(\omega)\mu_i(\omega)\frac{\xi^2}{c^2}}. \qquad (4)$$

Using proximity force approximation, the Casimir force between a sphere with a radius R and a plate can be approximately expressed by

$$F(a) = 2\pi R E(a), \tag{5}$$

where a denotes the minimum distance between the surface of the sphere and the plate. This approximation is available for $a \ll R$.

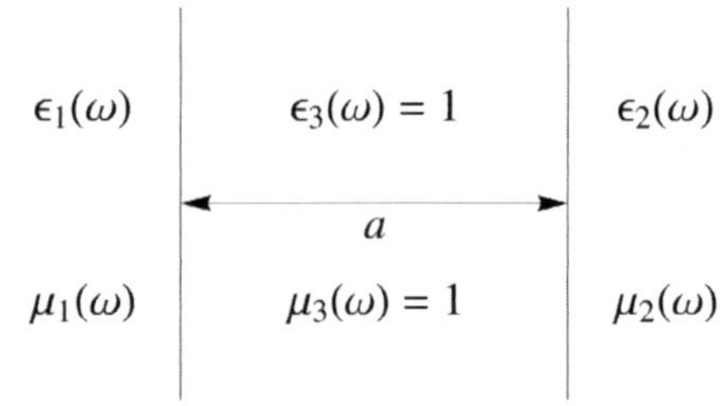

Figure 1. Illustration of configuration of a pair of infinite parallel plates. The space between plates is a vacuum.

2.2. Dielectric function of nickel

The Lifshitz formula requires the dielectric function and permeability along the imaginary frequency axis. Figure 2 shows the imaginary part of the dielectric function of nickel[17]. Since the experimental data lacks near the zero, we employ the Drude model[18],

$$\epsilon(\omega) = 1 - \frac{\omega_p^2}{\omega^2 + i\omega\omega_\tau}, \tag{6}$$

where $\omega_p = 7.42 \times 10^{15}$ (rad/s) and $\omega_\tau = 3.35 \times 10^{13}$ (rad/s).

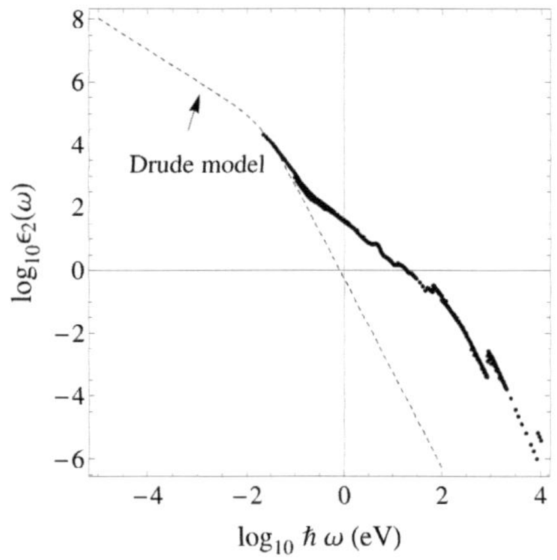

Figure 2. Plot of the imaginary component of the dielectric function of nickel. The dashing line denotes the extrapolation using the Drude model.

The dielectric function along the imaginary frequency axis can be obtained by the Kramers-Kronig relation

$$\epsilon(i\xi) = 1 + \frac{2}{\pi}\int_0^\infty dx \frac{x\mathrm{Im}\epsilon(x)}{x^2+\xi^2}. \tag{7}$$

Figure 3 shows the dielectric function of nickel and gold along the imaginary frequency axis[17]. We find that the shape of the dielectric function of nickel along the imaginary frequency axis is similar to that of gold.

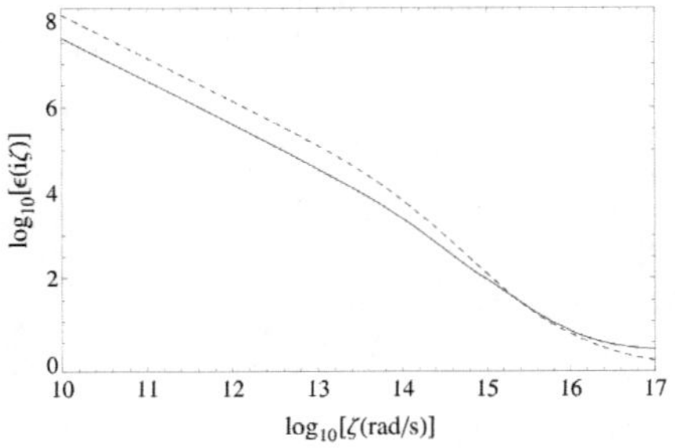

Figure 3. Plots of dielectric function along the imaginary frequency axis of gold (dashing line) and nickel (solid line).

2.3. Correction factors to the Casimir energy

Although nickel is a ferromagnetic material, we neglect the magnetic response of nickel and calculate the Casimir energy between a gold plate and a nickel plate. The Casimir energy strongly depends on the separation, we introduce a correction factor[1] is defined by

$$\eta_E = \frac{E(a)}{E_\infty(a)}, \tag{8}$$

where $E_\infty(a)$ is the Casimir energy between perfectly conductive plates per unit area. The Casimir energy between perfectly conductive plates[19] is given by

$$E_\infty(a) = -\frac{\pi^2\hbar c}{720a^3}. \tag{9}$$

Figure 4 shows the correction factor between a gold plate and a nickel plate (solid line) and that between a pair of gold plates (dashed line). Although the Casimir energy between gold plates is larger than that between a gold plate and a nickel plate, the difference between them becomes small as the separation decreases. As a result, if the magnetic response is negligible small, the dependence of the Casimir force between a gold sphere and a nickel plate is qualitatively the same with that between gold plates.

3. Casimir energy between a gold plate and a nanocomposite sheet

3.1. Nonmagnetic case

Let us consider a nanocomposite sheet containing many nickel nanoparticles. The electrical and magnetic response of a nanocomposite sheet is considerably different from those of a bulk nickel. At first, we focus on the change in the electrical property. It is not easy to express the dielectric

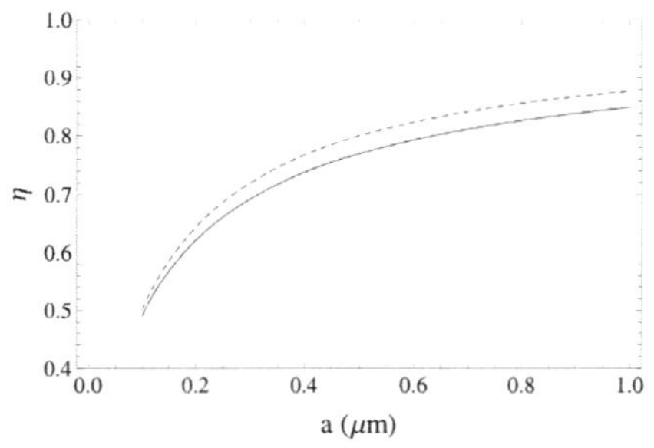

Figure 4. Plots of the correction factor of the Casimir energy between a gold plate and a nickel plate.

function of a nanocomposite sheet accurately. We assume that the dielectric function $\epsilon_{\rm NC}$ can be expressed by Maxwell-Garnett mixing formula as

$$\epsilon_{\rm NC} = \frac{\epsilon_m\left[(1+2f)\epsilon_{\rm Ni} + 2(1-f)\epsilon_m\right]}{(1-f)\epsilon_{\rm Ni} + (2+f)\epsilon_m}, \tag{10}$$

where ϵ_m denotes the dielectric function of a matrix and f denotes volume fraction. In the nanocomposite synthesized by Akamatsu et al. polyimide film was used as a matrix. Unfortunately, the experimental data of the dielectric function of polyimide film is very few. However, the value of the dielectric function of polyimide film is much smaller than that of nickel. Therefore, we assume that $\epsilon_m = 1$ for any frequency. Akamatsu et al. had succeeded to change volume fraction of nickel naoparticles up to 30%.

We neglect the magnetic response of the nanocomposite and show the correction factors of the Casimir energy between a gold plate and nanocomposite sheet including different concentration of nickel in figure 5. The correction factions become small by comparison with that between a gold plate and a bulk nickel plate.

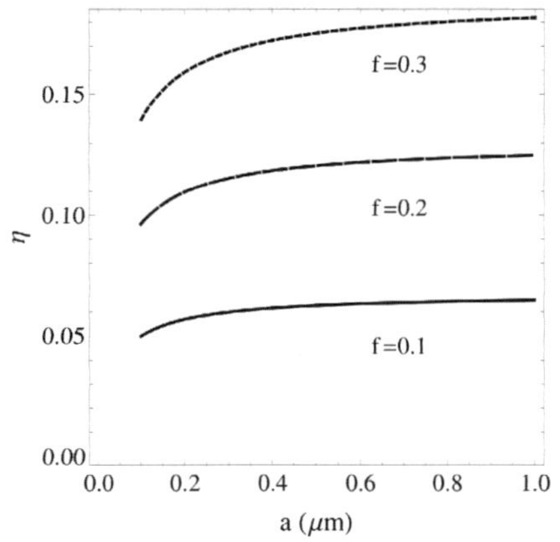

Figure 5. Plots of the correction factor of the Casimir energy between a gold plate and a nanocomposite sheet including nanoparticle with volume concentration f, in which the magnetic response is neglected.

3.2. Magnetic case

We now take the magnetic response of nickel into account. According to Ref. [13] we use a Debye relaxation model to express the permeability as

$$\mu_{\rm NC} = \mu_\infty + \frac{\Delta}{1 - i\omega\tau_c}, \tag{11}$$

where $\mu_\infty = 0.7333$, $\Delta = 9.98$, and $\tau_c^{-1} = 1.78 \times 10^{12}$ Hz. Figure 6 shows the magnetic permeability along the imaginary frequency axis for different volume fractions. Substituting

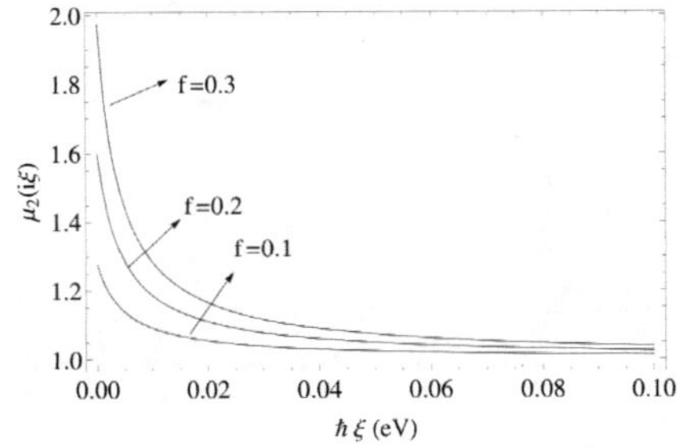

Figure 6. Plots of magnetic permeability of nanocomposite along the imaginary frequency axis with volume fraction f.

the obtained dielectric function and the magnetic permeability along the imaginary frequency axis into the Lifshitz formula we obtain the correction factor of the Casimir energy between a gold plate and a nanocomposite sheet in figure 7. The most significant difference between the correction of the Casimir energy for the system in which the magnetic response correctly and that in which the magnetic response is neglected is that the maximum exists for the former case below 1 μm separation. We find a maximum for every panel in figure 7. We note that the Casimir energy itself monotonically decreases as the separation increases. However, the maximum appears by normalized by the Casimir energy between perfectly conductive plates.

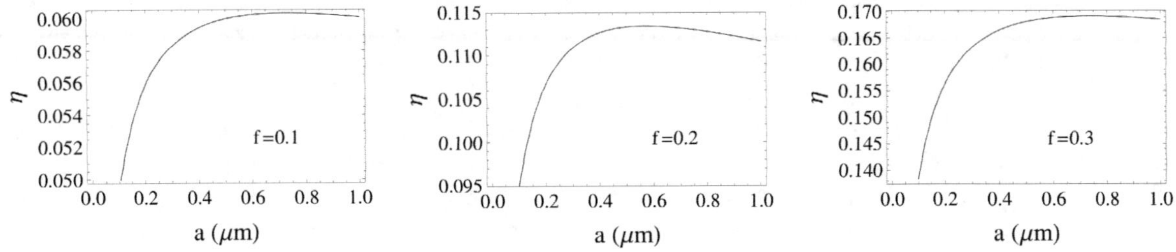

Figure 7. Plots of the correction factor of the Casimir energy between a gold plate and nanocomposite including nanoparticles whose the magnetic response is taken account in.

The decrease in the correction of the Casimir energy becomes pronounced for large separation. Figure 8 shows the correction of the Casimir energy from $a = 1$ μm to $a = 25$ μ m. We find the correction factor of the Casimir energy for $f = 0.1$ becomes negative near $a = 13\ \mu$ m. Since the Casimir force between a sphere and a plate is given by eq. (5), the Casimir force between a gold sphere and a nanocomposite sheet becomes repulsive for the negative correction factor of the Casimir energy. As a result, it is conjectured that the Casimir force between a gold sphere

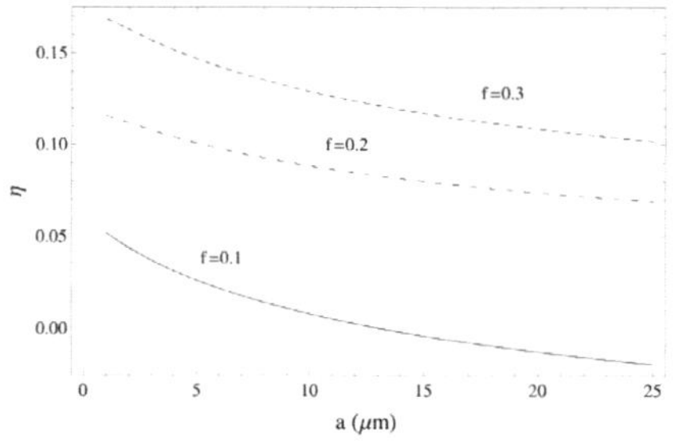

Figure 8. Plots of the correction factor of the Casimir energy between a gold plate and nanocomposite for large separations. The sign of the correction factor corresponding to $f = 0.1$ changes from positive to negative.

and a nanocomposite consisted of nickel nanoparticles becomes repulsive for large separations.

4. Conclusion

Metamaterial is a promising candidate that enables to levitate by the repulsive Casimir force[21]. However, if we choose a nanocomposite sheet as metamaterial, we have to strengthen the repulsive force acting on the nanocomposite sheet. Because the separation at which the Casimir force change from attractive to repulsive is much large, and it implies the repulsive Casimir force is very weak. As mentioned in earlier studies, it is difficult to generate the strong repulsive Casimir force when the dielectric property and magnetic property coexist. Nickel has a large imaginary dielectric component. Thus it hinders in generating the strong repulsive Casimir force, which results from magnetic effect. Nevertheless nanocomposite that exhibits magnetic response is valuable material for the investigations of the Casimir effect. Because we know little about the Casimir force between a dielectric object and a magnetic object.

An advantage of the nanocomposite over other metamaterials is the simplicity of structure. Many mematerials need complicated structure in order to exhibit a magnetic response. This implies that controlling the separation between a prove such as a gold sphere and a surface of metamaterial with complicated structure is difficult. On the other hand, nanocomposite including nickel nanopartic1 has already synthesized and it is possible to make its surface smooth. Accordingly nanocomposite sheet may be good material to test whether the Lifshitz formula is available for magnetic objects.

A simple but not rigorous test is to detect the maximum of the correction factor of the Casimir energy conjectured in this paper experimentally. The Casimir energy between perfectly conductive plates decreases in proportional to a^{-3} [20]. If the conductivities of the plates are finite and they are described by plasma model, the Casimir force decreases in proportional to a^{-2} for short separations and a^{-3} for large separations. In figure 5 the slope of the correction factor becomes small for large separations, and it indicates that the Casimir energy decreases in proportional to a^{-3} for large separations. On other hand, the slope of the correction factor of the Casimir energy in figure 7 increases for larger separation. This behavior is a unique property that cannot be explained by a simple finite-conductive model, and therefore it will be a touchstone.

Acknowledgment
This research was financially supported by the Foundation of the Shin Meiwa Kawanishi for the Education.

References

[1] Bordag M, Geyer B, Klimchitskaya G L, Mohideen U, and Mostepanenko V M 2009 *Advances in the Casimir Effect* (Oxford University Press, New York)
[2] Feiler A A, Bergström L, and Rutland N W 2008 it Langmuir **24** 2274
[3] Munday J N, Cappasso F and Parsegian V A 2009 *Nature* **457** 170
[4] Kenneth O and Klich I 2006 *Phys. Rev. Lett.* **97** 16041
[5] Genet C, Lambrecht A, and Reynaud S 2003 *Phys. Rev. A* **67** 43811
[6] Geyer B. Klimchitskaya G L, Mostepanenko V M 2010 *Phys. Rev. B* **81** 104101
[7] Rodriguez A W, Joannopoulos J D, and Johson S G 2008 *Phys. Rev. Lett* **77** 062107
[8] Boyer T H 1974 *Phys. Rev. A* **9** 2078
[9] Veselago V, Bragnisky L, Shklover V, and Hafner C 2006 *J. Computational and Theoretical Nanoscience* **3** 1
[10] Rosa F S S, Dalvit D A R, and Milonni P W 2008 *Phys. Rev. Lett.* **100** 183602
[11] Pirozhenko I G and Lambrecht A 2008 *J. Phys. A* **41** 164015
[12] Rosa F S S, Dalvit D A R, and Milonni P W 2008 *Phys. Rev. A* **78** 032117
[13] Yannopapas V and Vitanov N V 2009 *Phys. Rev. Lett.* **103** 120401
[14] Akamatsu K, Adachi S, Tsuoka T, Ikeda S, Tomita S, and Nawafune H 2008 *Chem. Mater.* **20** 3042
[15] Landau L D and Lifshitz E M, 1982 *Electrodynamics of Continuous Media* (Oxford Science Publications, SanDiego)
[16] Israelachvili J N 1992 *Intermolecular and Surface Forces* (Academic Press, London).
[17] Palik E D 1991 *Handbook of Optical Constants of Solids* (Academic, New York)
[18] Ordal M A, Long L L, Bell R J, Bell S E, Bell R R, Alexander R W Jr, and Ward C A 1983 *App. Optics* **22** 1099
[19] Casimir H B G 1948 *Proc. Kon. Ned. Akad. Wet.* **51** 793
[20] Mionni P W 1994 *The Quantum Vacuum* (Academic Press, SanDiego)
[21] Inui N and Miura K 2010 *eJSSN* **8** 57

Geometric Approach to Quantum Statistical Mechanics and Application to Casimir Energy and Friction Properties

Shoichi Ichinose

Laboratory of Physics, School of Food and Nutritional Sciences, University of Shizuoka
Yada 52-1, Shizuoka 422-8526, Japan

E-mail: ichinose@u-shizuoka-ken.ac.jp

Abstract. A geometric approach to general quantum statistical systems (including the harmonic oscillator) is presented. It is applied to Casimir energy and the dissipative system with friction. We regard the (N+1)-dimensional Euclidean *coordinate* system (X^i,τ) as the quantum statistical system of N quantum (statistical) variables (X^i) and one *Euclidean time* variable (τ). Introducing paths (lines or hypersurfaces) in this space (X^i,τ), we adopt the path-integral method to quantize the mechanical system. This is a new view of (statistical) quantization of the *mechanical* system. The system Hamiltonian appears as the *area*. We show quantization is realized by the *minimal area principle* in the present geometric approach. When we take a *line* as the path, the path-integral expressions of the free energy are shown to be the ordinary ones (such as N harmonic oscillators) or their simple variation. When we take a *hyper-surface* as the path, the system Hamiltonian is given by the *area* of the *hyper-surface* which is defined as a *closed-string configuration* in the bulk space. In this case, the system becomes a O(N) non-linear model. We show the recently-proposed 5 dimensional Casimir energy (ArXiv:0801.3064,0812.1263) is valid. We apply this approach to the visco-elastic system, and present a new method using the path-integral for the calculation of the dissipative properties.

1. Introduction

In the quest for the fundamental structure of the space, time, and matter, the most advanced theories are the string theory, D-brane theory and M-theory[1]. They are beyond the quantum field theory in that the extended (in space-time) objects are treated as fundamental elements. Since the finding of AdS/CFT correspondence[2, 3, 4], various new ideas and techniques, developed for them, are imported into the *non-perturbative* analysis of the quantum field theories. In particular, the application to the material physics is marvelous: heavy ion collision physics and the viscosity in the quark-gluon plasma([5, 6, 7] for review), superconductivity and superfluidity [8, 9, 10, 11], baryon mass spectrum in QCD[12, 13]. In this circumstance, a *new standpoint* about the space-time quantization appear. One is proposed by Hořava[14, 15]. He introduced Lifshitz's higher-derivative scalar theory and its renormalization group behavior into his idea about the new quantum gravity. Another one is revively given by E. Verlinde[16]. He emphasizes the entropic force (rather than energetic force) and the thermodynamical behavior near the horizon (Hawking radiation). With this recent trend of the geometrical view, the statistical(thermal) view and the visco-elastic view , we present a *new* formalism where the quantum statistical system is treated purely in the geometrical way.

In the analysis of the dissipative material-system, due to friction, one familiar way is to treat it as the *hydrodynamic* system of a continuum medium. The incompressible viscous flow (Newton' flow is assumed) is discribed by Navier-Stokes equation.

$$\frac{Dv^i}{Dt} \equiv (\frac{\partial}{\partial t} + v^j\partial_j)v^i = -\frac{1}{\rho}\partial_i P + \frac{\eta}{\rho}\Delta v^i + g^i \quad ,$$
$$i,j = 1,2,3 \quad , \quad \vec{x} \equiv (x^1, x^2, x^3) = (x,y,z) \quad , \quad \partial_i \equiv \frac{\partial}{\partial x^i} \quad , \quad \Delta \equiv \partial_1^2 + \partial_2^2 + \partial_3^2 \quad ,$$
$$v^i = v^i(t,\vec{x}) \quad , \quad P = P(t,\vec{x}) \quad , \qquad (1)$$

where $\vec{g} \equiv (0,0,g)$ is the gravitational acceleration constant. ρ and η is the mass density (constant) and the viscosity (constant) of the fluid, respectively. $\vec{v} \equiv (v^1, v^2, v^3)$ is the velocity field and $P(t,\vec{x})$ is the pressure field. This equation has been examined analytically and numericallly in various ways. The statistical quantities of the system has been obtained, for example, by introducing the random force (Langevin equation, the stochastic method, Fokker-Planck equation). In the present work, by introducing a *metric* in (t,x^i)-space, we treat this system geometrically. The statistical or global quantities (energy, entropy,...) are exprssed in the *path-integral* form.

As another application, we take Casimir energy. It is the vcuum (zero temperature) energy for the free (non-interacting) part of the quantum system. It is a basic quantity of the quantum field theory (QFT) such as the field of radiation (the electromagnetic field). Generally it is expressed by the boundary parameters of the system. The quantity is so delicately defined that we need careful regularization of divergences. The present approach gives a new regularization which has some characteristic points compared with the ordinary one. Especially the direction of the *renormalization group* flow determines the attractiveness or repulsiveness. We also apply the present approach to 5 dimensional (dim) field theories (flat and curved ones), and well-define the higher-dimensional QFTs.

The content is organized as follows. We start with the simple quantum statistical system of one harmonic oscillator in Sec.2. We see the geometric approach works well by regarding the *extra coordinate* as the *Euclidean time*. We generalize the harmonic oscillator potential (elastic system) to the general one in Sec.3. In Sec.4 the one variable system is generalized to the system of N variables. We analyze the quantum statistical system in the N+1 extra dimensional Euclidean *geometry*. The O(N) nonliner model naturally appears by taking the *closed-string* configuration. We stress that taking the area as Hamiltonian is one realization of the *minimal area principle*. In Sec.5, we apply the present approach to Casimir energy. The visco-elastic system is treated in Sec.6. We conclude in Sec.7. In Appendix A, some detail of Sec.5.2 (Casimir energy in 5D curved space) is explained.

2. Quantum Statistical System of Harmonic Oscillator

2.1. 'Dirac' Type

Let us consider 2 dim Euclidean space (X,τ) described by the following metric.

$$ds^2 = dX^2 + \omega^2 X^2 d\tau^2 = G_{AB}dX^A dX^B \quad ,$$
$$(X^A) = (X^1, X^2) = (X,\tau) \quad , \quad (G_{AB}) = \mathrm{diag}(1, \omega^2 X^2) \quad ,$$
$$R_{AB} = 0 \quad , \quad R = G^{AB}R_{AB} = 0 \quad , \qquad (2)$$

where $A, B = 1, 2$. The parameter ω is the 'spring' constant with the dimension of mass. We impose the periodicity (period: β) in the direction of the extra dimension τ.

$$\tau \to \tau + \beta \quad . \qquad (3)$$

This is a way to introduce the *temperature* $(1/\beta)$ in the system. Here we take a path $\{x(\tau),\ 0 \leq \tau \leq \beta\}$ in the 2D bulk space (X,τ) and the *induced* metric on the *line* is given by

$$X = x(\tau) \quad , \quad dX = \dot{x}d\tau \quad , \quad \dot{x} \equiv \frac{dx}{d\tau} \quad , \quad 0 \leq \tau \leq \beta \quad ,$$
$$ds^2 = (\dot{x}^2 + \omega^2 x^2)d\tau^2 \quad . \tag{4}$$

See Fig.1.

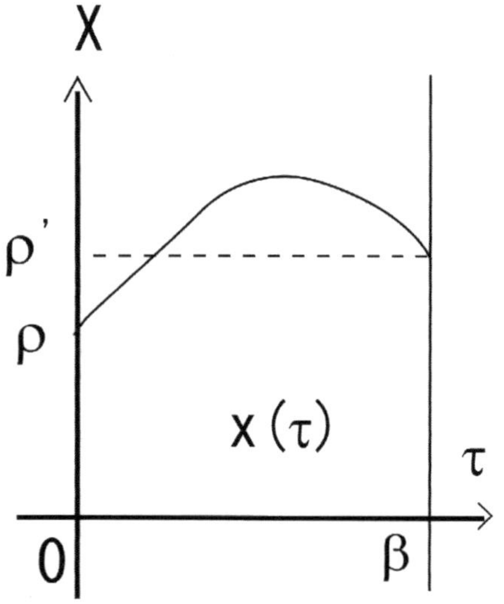

Figure 1. A path of line in 2D Euclidean space (X,τ). The path starts at x(0)=ρ and ends at x(β)=ρ'.

Then the *length* L of the path $x(\tau)$ is given by

$$L = \int ds = \int_0^\beta \sqrt{\dot{x}^2 + \omega^2 x^2} d\tau \quad . \tag{5}$$

We take the half of the length $(\frac{1}{2}L)$ as the system Hamiltonian (*minimal length principle*). Then the free energy F of the system is given by

$$\mathrm{e}^{-\beta F} = \int_{-\infty}^{\infty} d\rho \int_{\substack{x(0) = \rho \\ x(\beta) = \rho}} \prod_\tau \mathcal{D}x(\tau) \exp\left[-\frac{1}{2} \int_0^\beta \sqrt{\dot{x}^2 + \omega^2 x^2} d\tau \right] \quad , \tag{6}$$

where the path-integral is done for all possible paths with the indicated boundary condition (b.c.). This quantum statistical system can be regarded as the square-root type ('Dirac' type) of the ordinary harmonic oscillator. [1]

2.2. Standard Type

Now we consider another type of 2 dim Euclidean space (X, τ) described by the following *line* element.

$$ds^2 = \frac{1}{d\tau^2}(dX^2)^2 + \omega^4 X^4 d\tau^2 + 2\omega^2 X^2 dX^2 = \frac{1}{d\tau^2}(dX^2 + \omega^2 X^2 d\tau^2)^2 \quad , \tag{7}$$

[1] The situation reminds us of the relation between Nambu-Goto action and Polyakov action in the string theory[17]. The introduction of an auxiliary variable helps to 'normalize' the square-root action (6). In this case, the geometric role of the auxiliary variable remains obscure.

where we put the following condition on the infinitesimal quantities, $d\tau^2$ and dX^2, in order to keep all terms of (7) in the same order.

$$\text{[Line Element Regularity Condition]} : \quad d\tau^2 \sim O(\epsilon^2) \quad , \quad dX^2 \sim O(\epsilon^2) \quad , \quad \frac{1}{d\tau^2}dX^2 \sim O(1) \quad , \tag{8}$$

where ϵ is an arbitrary infinitesimal parameter with the dimension of length. [2] Note that we do *not* have 2D metric in this case. (We cannot define the bulk metric $G_{AB}(X)$.) We impose the periodicity (period: β).

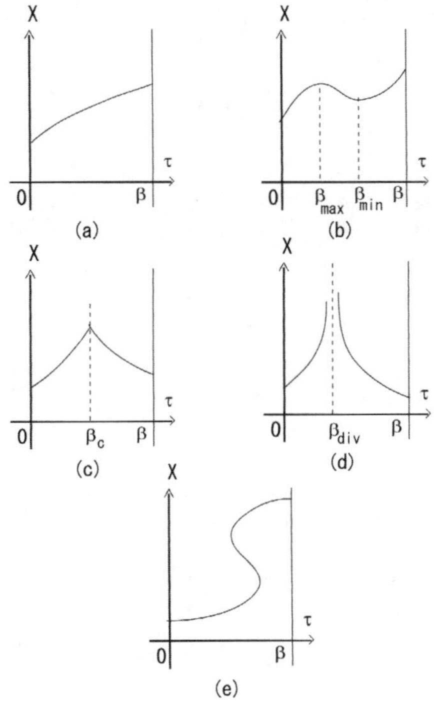

Figure 2. Singular and regular lines in 2D Euclidean space (X,τ). (a) regular line, simply increasing; (b) regular line, maximum at β_{max} and minimum at β_{min}; (c) singular line, different derivatives for $\beta \rightarrow \beta_c \pm 0$; (d) singular line, divergent at β_{div}; (e) singular line, multi-valued.

$$\tau \rightarrow \tau + \beta \quad . \tag{9}$$

Here we take a path $\{x(\tau),\ 0 \leq \tau \leq \beta\}$, and the *induced* metric on the line is given by

$$X = x(\tau) \quad , \quad dX = \dot{x}d\tau \quad , \quad \dot{x} \equiv \frac{dx}{d\tau} \quad , \quad 0 \leq \tau \leq \beta \quad ,$$
$$ds^2 = (\dot{x}^2 + \omega^2 x^2)^2 d\tau^2 \quad . \tag{10}$$

In the bulk we do *not* have the metric, but on the path, we *do* have this *induced* metric. Then the *length* L of the path $x(\tau)$ is given by

$$L[x(\tau)] = \int ds = \int_0^\beta (\dot{x}^2 + \omega^2 x^2) d\tau \quad . \tag{11}$$

[2] The condition (8) restricts the trajectory configuration (10) only to smooth-lines in the 2D bulk space, and excludes singular-lines which have some *singular* points (the derivative along τ can not be defined) between $0 \leq \tau \leq \beta$. See Fig.2 for singular and regular lines.

Hence, taking $\frac{1}{2}L$ as the Hamiltonian (*minimal length principle*), the free energy F of the system is given by

$$\mathrm{e}^{-\beta F} = \int_{-\infty}^{\infty} d\rho \int_{\substack{x(0)=\rho \\ x(\beta)=\rho}} \prod_{\tau} \mathcal{D}x(\tau) \exp\left[-\frac{1}{2}\int_0^\beta (\dot{x}^2 + \omega^2 x^2) d\tau \right] \quad , \tag{12}$$

where the path-integral is done for all possible paths with the indicated b.c.. This is exactly the free energy of the harmonic oscillator. See Feynman's textbook[18]. [3]

Note that the condition (8) is necessary for the *elastic* view to the path.

3. General Quantum Statistical System

We generalize the harmonic oscillator potential, $\frac{1}{2}\omega^2 X^2$, to the general one $V(X)$. As for $V(X)$, we have the following form in mind.

$$\frac{\omega^2}{2}X^2 + \frac{\lambda_3}{3!}X^3 + \frac{\lambda_4}{4!}X^4 + \cdots \quad , \tag{13}$$

where $\lambda_3, \lambda_4, \cdots$ are the coupling constants for additional terms.

3.1. 'Dirac' Type

We start with the following metric in 2 dim Euclidean space (X, τ).

$$\begin{gathered} ds^2 = dX^2 + 2V(X)d\tau^2 = G_{AB}dX^A dX^B \quad , \\ (X^A) = (X^1, X^2) = (X, \tau) \quad , \quad (G_{AB}) = \mathrm{diag}(1, 2V(X)) \quad , \\ (R_{AB}) = \begin{pmatrix} \frac{V''}{2V} - \frac{1}{4}\left(\frac{V'}{V}\right)^2, & 0 \\ 0 \;, & V'' - \frac{1}{2}\frac{(V')^2}{V} \end{pmatrix} \quad , \quad R = G^{AB}R_{AB} = \frac{V''}{V} - \frac{1}{2}\left(\frac{V'}{V}\right)^2 \quad , \\ V' \equiv \frac{dV(X)}{dX} \quad , \quad V'' \equiv \frac{d^2V(X)}{dX^2} \quad , \end{gathered} \tag{14}$$

where $A, B = 1, 2$. Note that V(X) does *not* depend on τ. We impose the periodicity (period: β) in the direction of the extra dimension τ (3). On a path $\{x(\tau),\ 0 \leq \tau \leq \beta\}$, the *induced* metric is given by

$$ds^2 = (\dot{x}^2 + 2V(x))d\tau^2 \quad , \quad 0 \leq \tau \leq \beta \quad . \tag{15}$$

Hence the *length* L of the path $x(\tau)$ is given by

$$L = \int ds = \int_0^\beta \sqrt{\dot{x}^2 + 2V(x)}d\tau \quad . \tag{16}$$

Taking the half of the length ($\frac{1}{2}L$) as the Hamiltonian, we get the free energy F as

$$\mathrm{e}^{-\beta F} = \int_{-\infty}^{\infty} d\rho \int_{\substack{x(0)=\rho \\ x(\beta)=\rho}} \prod_{\tau} \mathcal{D}x(\tau) \exp\left[-\frac{1}{2}\int_0^\beta \sqrt{\dot{x}^2 + 2V(x)}d\tau \right] \quad . \tag{17}$$

[3] $F = \frac{\omega}{2} + \frac{1}{\beta}\ln(1-\mathrm{e}^{-\beta\omega}), E = <\frac{L}{2}> = \frac{\omega}{2}\coth(\frac{\omega\beta}{2}) = \frac{\omega}{2} + \frac{\omega}{\mathrm{e}^{\omega\beta}-1}, S = \frac{1}{T}(E-F) = k\{\frac{\beta\omega}{2}\coth\frac{\beta\omega}{2} - \frac{\beta\omega}{2} - \ln(1-\mathrm{e}^{-\beta\omega})\}$

3.2. Standard Type

We start with the following line element.

$$ds^2 = \frac{1}{d\tau^2}(dX^2)^2 + 4V(X)^2 d\tau^2 + 4V(X)dX^2 = \frac{1}{d\tau^2}\left(dX^2 + 2V(X)d\tau^2\right)^2 \quad , \tag{18}$$

where we put the condition (8) on the infinitesimal quantities, $d\tau^2$ and dX^2, in order to keep all terms in the same order. The 2D bulk space do *not* have 2D metric. We impose the periodicity (period: β) (9). On a path $\{x(\tau),\ 0 \leq \tau \leq \beta\}$, we have the *induced* metric:

$$ds^2 = (\dot{x}^2 + 2V(x))^2 d\tau^2 \quad . \tag{19}$$

The *length* L is given by

$$L[x(\tau)] = \int ds = \int_0^\beta (\dot{x}^2 + 2V(x))d\tau \quad . \tag{20}$$

Taking $\frac{1}{2}L$ as the Hamiltonian, the free energy F is given by

$$\mathrm{e}^{-\beta F} = \int_{-\infty}^{\infty} d\rho \int_{\substack{x(0)=\rho \\ x(\beta)=\rho}} \prod_\tau \mathcal{D}x(\tau) \exp\left[-\frac{1}{2}\int_0^\beta (\dot{x}^2 + 2V(x))d\tau\right] \quad , \tag{21}$$

where the path-integral is done for all possible paths with the indicated b.c.. This is exactly the free energy of the quantum statistical system of one variable x in the general potential V(x).

4. Quantum Statistical System of N Harmonic Oscillators and O(N) Nonlinear Model

4.1. 'Dirac' Type of N Harmonic Oscillators and O(N) nonlinear system

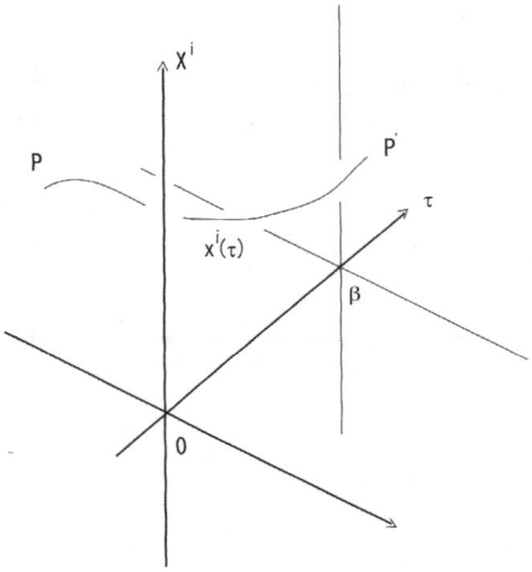

Figure 3. A path of line $\{x^i(\tau)|i = 1,2,\cdots,N\}$ in N(=2)+1 dim space. It starts at P=$(\rho_1,\rho_2,\cdots,\rho_N,0)$ and ends at P$'$=$(\rho'_1,\rho'_2,\cdots,\rho'_N,\beta)$.

Let us consider N+1 dim Euclidean space $(X^i,\tau), i = 1,2,\cdots,N$ described by the following metric.

$$ds^2 = \sum_{i=1}^N (dX^i)^2 + \omega^2 d\tau^2 \sum_{i=1}^N (X^i)^2 = \sum_{i=1}^N (dX^i)^2 + \omega^2 r^2 d\tau^2 = G_{AB}dX^A dX^B \quad ,$$
$$A, B = 1,2,\cdots,N,N+1; \quad X^{N+1} \equiv \tau \quad ,$$
$$(G_{AB}) = \mathrm{diag}(1,1,\cdots,1,\omega^2 r^2) \quad , \quad r^2 \equiv \sum_{i=1}^N (X^i)^2 \quad . \tag{22}$$

(Subsec.2.1 is the $N = 1$ case.) The Ricci tensor and the scalar curvature are, for N=2, given by [4]

$$ds^2 = dx^2 + dy^2 + \omega^2(x^2+y^2)d\tau^2 \quad ,$$
$$(R_{AB}) = \frac{1}{(r^2)^2}\begin{pmatrix} y^2 & -xy & 0 \\ -yx & x^2 & 0 \\ 0 & 0 & \omega^2(r^2)^2 \end{pmatrix} \quad , \quad R = \frac{2}{r^2} > 0 \quad , \quad r^2 = x^2 + y^2 \quad ,$$
$$\sqrt{G} = \omega\sqrt{x^2+y^2} \quad , \quad \sqrt{G}R = \frac{2\omega}{\sqrt{x^2+y^2}} \tag{23}$$

where $(X^1, X^2, X^3) = (x, y, \tau)$ is taken.

We impose the periodicity (3)(period: β), and take a path $\{X^i = x^i(\tau)|\ 0 \leq \tau \leq \beta,\ i = 1, 2, \cdots, N\}$(See Fig.3). The *induced* metric on the *line* is given by

$$X^i = x^i(\tau) \quad , \quad dX^i = \dot{x}^i d\tau \quad , \quad \dot{x}^i \equiv \frac{dx^i}{d\tau} \quad , \quad 0 \leq \tau \leq \beta \quad ,$$
$$i = 1, 2, \cdots, N \quad , \quad ds^2 = \sum_{i=1}^{N}((\dot{x}^i)^2 + \omega^2(x^i)^2)d\tau^2 \quad . \tag{24}$$

Then the *length* L of the path $\{x^i(\tau)\}$ is given by

$$L = \int ds = \int_0^\beta \sqrt{\sum_{i=1}^{N}((\dot{x}^i)^2 + \omega^2(x^i)^2)}\ d\tau \quad . \tag{25}$$

We take the half of the length ($\frac{1}{2}L$) as the system Hamiltonian(*minimal length principle*). Then the free energy F of the system is given by

$$\mathrm{e}^{-\beta F} = (\prod_i \int_{-\infty}^{\infty} d\rho_i) \int_{\substack{x^i(0)=\rho_i \\ x^i(\beta)=\rho_i}} \prod_{\tau,i} \mathcal{D}x^i(\tau) \exp\left[-\frac{1}{2}\int_0^\beta \sqrt{\sum_{i=1}^{N}((\dot{x}^i)^2 + \omega^2 x^{i2})d\tau}\right] \quad , \tag{26}$$

where the path-integral is done for all possible paths $\{x^i(\tau); i = 1, 2, \cdots N\}$ with the indicated b.c.. We can regard this as the free energy for a variation ('Dirac' type) of the N harmonic oscillators's. (See next subsection for the ordinary type of the N harmonic oscillators.)

Instead of the length L, we can take another geometric quantity. Let us consider the following N dim *hypersurface* in N+1 dim space (a closed-string configuration). See Fig.4 for the N=2 case.

$$\sum_{i=1}^{N}(X^i)^2 = r^2(\tau) \quad , \quad \sum_{i=1}^{N} X^i dX^i = r\dot{r}d\tau \quad , \quad 0 \leq \tau \leq \beta \quad . \tag{27}$$

The form of $r(\tau)$ describes a path (N dimensional hypersurface in the bulk) which is *isotropic* in the 'brane' at τ (the N dim space 'perpendicularly' standing at τ of the extra axis, not the hypersurface). The *induced* metric on the N dim hypersurface is given by

$$ds^2 = \sum_{i,j}(\delta_{ij} + \frac{\omega^2}{\dot{r}^2}x^i x^j)dx^i dx^j \equiv \sum_{i,j} g_{ij}dx^i dx^j \quad ,$$
$$g_{ij} = \delta_{ij} + \frac{\omega^2}{\dot{r}^2}x^i x^j \quad , \quad r^2 = \sum_{i=1}^{N}(x^i)^2 \quad , \quad \det(g_{ij}) = 1 + \frac{\omega^2 r^2}{\dot{r}^2} \quad . \tag{28}$$

[4] All curvature calculation in this work is checked by the algebraic calculation soft "Maxima"[19].

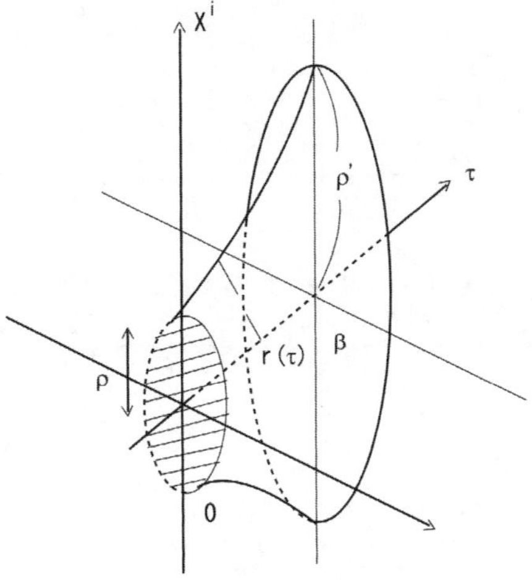

Figure 4. A path of hyper-surface. N(=2) dim hypersurface in N+1 dim space $(X^1, X^2, \cdots, X^N, \tau)$. S^{N-1} radius $r(\tau)$ starts with $r(0) = \rho$ and ends with $r(\beta) = \rho'$. We take this configuration as a path in the path integral (30) and (41). This is a closed-string configuration.

This is the metric of a O(N) nonlinear system and is the one dimensional *nonlinear sigma model* as the field theory. (The standard model (2 dim nonlinear sigma model) has often been used so far in order to show the *renormalization group* behavior of various systems. The background (effective action) formulation of the string theory heavily relies on the model.) Then the *area* of the N dim hypersurface is given by

$$A_N = \int \sqrt{\det g_{ij}}\, d^N x = \frac{N\pi^{N/2}}{\Gamma(\frac{N}{2}+1)} \int \sqrt{\dot{r}^2 + \omega^2 r^2} r^{N-1} d\tau \quad . \tag{29}$$

When we take $\frac{1}{2}A_N$ as the Hamiltonian (*minimal area principle*), the free energy F is given by

$$\mathrm{e}^{-\beta F} = \int_0^\infty d\rho \int_{\substack{r(0)=\rho \\ r(\beta)=\rho}} \prod_{\tau,i} \mathcal{D}x^i(\tau) \exp\left[-\frac{1}{2}\frac{N\pi^{N/2}}{\Gamma(\frac{N}{2}+1)} \int \sqrt{\dot{r}^2 + \omega^2 r^2} r^{N-1} d\tau \right] \quad . \tag{30}$$

We should compare this result ($N = 4$) with the proposed 5D Casimir energy for the *flat* geometry (59). The component $\sqrt{\dot{r}^2 + \omega^2 r^2}$ in the integrand of (30) is replaced by $\sqrt{\dot{r}^2 + 1}$ in (59).

We recognize, if we start with

$$ds^2 = \sum_{i=1}^{N} (dX^i)^2 + d\tau^2 \quad \text{(N+1 dim Euclidean flat)} \quad , \tag{31}$$

instead of (22), the integration measure becomes *exactly* the same as (59).

4.2. Standard Type of N Harmonic Oscillators

Now we consider another type of N+1 dim Euclidean space (X^i, τ); $i = 1, 2, \cdots N$ described by the following line element.

$$\begin{aligned} ds^2 &= d\tau^{-2}\{\sum_{i=1}^{N}(dX^i)^2\}^2 + \omega^4\{\sum_{i=1}^{N}(X^i)^2\}^2 d\tau^2 + 2\omega^2\{\sum_{i=1}^{N}(X^i)^2\}\{\sum_{j=1}^{N}(dX^j)^2\} \\ &= \frac{1}{d\tau^2}\{\sum_{i=1}^{N}(dX^i)^2 + \omega^2 r^2 d\tau^2\}^2 \quad , \quad r^2 = \sum_{i=1}^{N}(X^i)^2 \quad , \end{aligned} \tag{32}$$

with the condition:

$$\text{[Line Element Regularity Condition]} \ : \\ d\tau^2 \sim O(\epsilon^2) \quad , \quad (dX^i)^2 \sim O(\epsilon^2) \quad , \quad \frac{1}{d\tau^2}\{\sum_{i=1}^{N}(dX^i)^2\} \sim O(1) \ , \tag{33}$$

in order to keep all terms of (32) in the order of ϵ^2. [5] Again we note that, in the above case, we do *not* have N+1 dim (bulk) metric. We impose the periodicity (3): (period: β).

Here we take a path of Fig.3: $\{x^i(\tau)|\ 0 \leq \tau \leq \beta, i = 1, 2, \cdots, N\}$ and the *induced* metric on the path is given by

$$X^i = x^i(\tau) \quad , \quad dX^i = \dot{x}^i d\tau \quad , \quad \dot{x}^i \equiv \frac{dx^i}{d\tau} \quad , \quad 0 \leq \tau \leq \beta \quad , \\ ds^2 = [\sum_{i=1}^{N}((\dot{x}^i)^2 + \omega^2(x^i)^2)]^2 d\tau^2 \quad . \tag{34}$$

Then the *length* L of the path $\{x^i(\tau)\}$ is given by

$$L[x^i(\tau)] = \int ds = \int_0^\beta \sum_{i=1}^{N}((\dot{x}^i)^2 + \omega^2(x^i)^2)d\tau \quad . \tag{35}$$

Hence, taking $\frac{1}{2}L$ as the Hamiltonian (*minimal length principle*), the free energy F of the system is given by

$$\mathrm{e}^{-\beta F} = \left(\prod_i \int_{-\infty}^{\infty} d\rho_i\right) \int_{\substack{x^i(0) = \rho_i \\ x^i(\beta) = \rho_i}} \prod_{i,\tau} \mathcal{D}x^i(\tau) \exp\left[-\frac{1}{2}\int_0^\beta \sum_{i=1}^{N}((\dot{x}^i)^2 + \omega^2(x^i)^2)d\tau\right], \tag{36}$$

where the path-integral is done for all possible paths with the indicated b.c.. This is *exactly* the free energy of N harmonic oscillators.

We note again the condition (33) is necessary for the *elastic* view to the hyper-surfaces.

4.3. Middle type of O(N) nonlinear system

Instead of (32), we can start from a slightly modified metric.

$$ds^2 = \omega^4\{\sum_{i=1}^{N}(X^i)^2\}^2 d\tau^2 + 2\omega^2\kappa\{\sum_{i=1}^{N}(X^i)^2\}\{\sum_{j=1}^{N}(dX^j)^2\} \\ = \omega^2 r^2 \left(\omega^2 r^2 d\tau^2 + 2\kappa\sum_{j=1}^{N}(dX^j)^2\right) \quad , \quad r^2 = \sum_{i=1}^{N}(X^i)^2 \quad . \tag{37}$$

We drop the first term of (32), and add a free (real) parameter κ in the third one. We stress that, in this case, we need *not* the condition of (33). The line element is the ordinary type and

[5] As in (8), this condition restricts the trajectory configuration (34) only to *smooth* hyper-surfaces in the (N+1)-dim space.

we have the bulk metric G_{AB} in this case. The Ricci tensor and the scalar curvature, for $N=2$, are given by

$$ds^2 = \omega^4(x^2+y^2)^2 d\tau^2 + 2\omega^2\kappa(x^2+y^2)(dx^2+dy^2) \quad ,$$
$$(R_{AB}) = \frac{1}{(r^2)^2}\begin{pmatrix} 4y^2 & -4xy & 0 \\ -4yx & 4x^2 & 0 \\ 0 & 0 & \frac{2\omega^2}{\kappa}(r^2)^2 \end{pmatrix} \quad , \quad R = \frac{4}{\kappa\omega^2(r^2)^2} \quad , \quad r^2 = x^2+y^2 \quad ,$$
$$\sqrt{G} = 2\omega^4|\kappa|r^4 \quad , \quad \sqrt{G}R = 8\omega^2\cdot \mathrm{sign}(\kappa) \quad , \tag{38}$$

where $(X^1,X^2,X^3)=(x,y,\tau)$ and sign(κ) is the sign of κ. [6] We consider the N dim hypersurface (27), or Fig.4, and the *induced* metric on it is given by

$$ds^2 = \sum_{i,j=1}^{N} 2\omega^2 r^2(\kappa\delta_{ij} + \frac{1}{2}\frac{\omega^2}{\dot{r}^2}x^i x^j)dx^i dx^j \equiv \sum_{i,j} g_{ij}dx^i dx^j \quad . \tag{39}$$

Then the *area* of this hypersurface is given by

$$A_N = \int \sqrt{\det g_{ij}}\; d^N x = \frac{(2\pi\omega^2|\kappa|)^{N/2}}{\Gamma(\frac{N}{2}+1)}\int_0^\beta r^N\sqrt{\dot{r}^2 + \frac{r^2\omega^2}{2|\kappa|}}\; r^{N-1}d\tau \quad . \tag{40}$$

Taking $\frac{1}{2}A_N$ as the Hamiltonian (*minimal area principle*), the free energy, F, is given by

$$\mathrm{e}^{-\beta F} = \int_0^\infty d\rho \int_{\substack{r(0)=\rho \\ r(\beta)=\rho}} \prod_{\tau,i}\mathcal{D}x^i(\tau)\exp\left[-\frac{1}{2}\frac{(2\pi\omega^2|\kappa|)^{N/2}}{\Gamma(\frac{N}{2}+1)}\int_0^\beta r^N\sqrt{\dot{r}^2+\frac{r^2\omega^2}{2|\kappa|}}\; r^{N-1}d\tau\right] \quad .\tag{41}$$

We should compare this result (N=4,κ=1/2) with the proposed 5D Casimir energy for the *warped* geometry (59). They are similar ($(\omega r)^4\sqrt{\dot{r}^2+r^2\omega^2}$ of (41) is replaced by $(1/\omega z)^4\sqrt{r'^2+1}$ of (59).).

4.4. Modified type of O(N) nonlinear system
Instead of (37), we take the following type of metric.

$$ds^2 = W(\tau)\left(V(r)d\tau^2 + \sum_{j=1}^{N}(dX^j)^2\right) \quad , \quad r^2 = \sum_{i=1}^{N}(X^i)^2 \quad . \tag{42}$$

Especially, if we start with $W(\tau)=1/\tau^2, V(r)=1$,

$$\text{Euclidean (AdS)}_{N+1}: \quad ds^2 = \frac{1}{\tau^2}\{d\tau^2 + \sum_{j=1}^{N}(dX^j)^2\} \quad , \tag{43}$$

we recognize the integration measure *exactly* becomes the same as that in (59): $\tau^{-N}\sqrt{\dot{r}^2+1}r^{N-1}d\tau$.

The content in this section can be generalized for the *general* isotropic potential.

[6] $R>0$ for $\kappa>0$, $R<0$ for $\kappa<0$.

Journal of Physics: Conference Series **258** (2010) 012003 doi:10.1088/1742-6596/258/1/012003

5. Casimir Energy

One important application of the present approach is Casimir energy. As mentioned in the introduction, it requires carefull regularization to define the quantity rigorously. In order to deal with divergence properly, the IR and UV regularizations are important. The final (finite) result depends on the *boundary* parameters only.

5.1. 4D Flat Case (Ordinary Casimir Energy)

Before explaining the original motivation (the 5 dim Casimir energy), we discuss the ordinary case (1+3 dim electromagnetism) and the relation to the present approach. We consider the electromagnetism in Minkwski space:

$$ds^2 = -dt^2 + dx^2 + dy^2 + dz^2 \quad . \tag{44}$$

We place 2 perfectly-conducting plates parallel with the separation $2l$ in the x-direction. This configuration can be realized by taking the following boundary conditions. As for y- and z-directions, we impose the periodicity for the infra-red (IR) regularization.

$$\begin{aligned} \text{Periodicity}\ :\ x \to x + 2l \quad , \quad y \to y + 2L \quad , \quad z \to z + 2L \quad , \\ L \gg l \quad , \end{aligned} \tag{45}$$

where L is the IR-regularization parameter. Then the eigen frequencies of the electromagnetic wave and Casimir energy are given by

$$\begin{aligned} \omega_{n,m_y,m_z} &= \sqrt{(n\frac{\pi}{l})^2 + (m_y\frac{\pi}{L})^2 + (m_z\frac{\pi}{L})^2} \quad , \\ E_{Cas} &= 2 \cdot \sum_{n,m_y,m_z \in \mathbf{Z}} \frac{1}{2}\omega_{n,m_y,m_z} \quad \geq 0 \quad , \end{aligned} \tag{46}$$

where $\mathbf{Z}$ indicates all integers. $\frac{1}{2}\omega_{n,m_y,m_z}$ is the zero-point oscillation energy. Introducing the cut-off function $g(x)$ ($= 1$ for $0 < x < 1$, 0 for otherwise), Casimir energy can formally be written as

$$E^{\Lambda}_{Cas} = \sum_{n,m_y,m_z \in \mathbf{Z}} \omega_{n,m_y,m_z} g\left(\frac{\omega_{n,m_y,m_z}}{\Lambda}\right) \quad \geq 0 \quad . \tag{47}$$

We take the continuum limit $L \to \infty$, $L \ll l \to \infty$.

$$\begin{aligned} E^{\Lambda 0}_{Cas} &= \int_{-\infty}^{\infty}\int_{-\infty}^{\infty} \frac{dk_y dk_z}{(\frac{\pi}{L})^2} \int_{-\infty}^{\infty} \frac{dk_x}{\frac{\pi}{l}} \sqrt{k_x^2 + k_y^2 + k_z^2}\, g(\frac{k}{\Lambda}) \\ &= \int_{-\infty}^{\infty}\int_{-\infty}^{\infty}\int_{-\infty\,|k|\leq\Lambda}^{\infty} \frac{dk_x dk_y dk_z}{(\frac{\pi}{L})^2 \frac{\pi}{l}} \sqrt{k_x^2 + k_y^2 + k_z^2} \quad \geq 0 \quad . \end{aligned} \tag{48}$$

Note that E_{Cas}, E^{Λ}_{Cas} and $E^{\Lambda 0}_{Cas}$ are all positive-definite.

In a familiar way, regarding $E^{\Lambda 0}_{Cas}$ as the origin of the energy scale, we consider the quantity $u = (E^{\Lambda}_{Cas} - E^{\Lambda 0}_{Cas})/(2L)^2$ as the physical Casimir energy and evaluate it with the help of the Euler-MacLaurin formula as $u = (\pi^2/(2l)^3)\ (B_4/4!) = -(\pi^2/720)(1/(2l)^3) < 0$. [7] The final result is negative. In the present analysis we take a new regularization.

[7] $B_4 = -1/30$ is the 4-th Bernoulli number.

First we re-express $E^{\Lambda 0}_{Cas}$ using a simple identity : $l = \int_0^l dw$ (w: a regularization or 'extra' axis).

$$
E^{\Lambda 0}_{Cas}/(2L)^2 = \frac{1}{2^2\pi^3}\int_0^l dw \int_{k\leq\Lambda} P(k) 2\pi k^2 dk
$$
$$
= \frac{1}{2^2\pi^3}\int_0^l dw(-1)\int_{r\geq\Lambda^{-1}} P(1/r)(-1)2\pi r^{-4} dr \quad .
$$
$$
P(k) \equiv k \quad , \quad r \equiv \frac{1}{k} \quad , \tag{49}
$$

where the integration variable changes from the momentum (k) to the coordinate ($r = \sqrt{x^2+y^2+z^2}$). The integration region in (R, w)-space is the infinite rectangular shown in Fig.5.

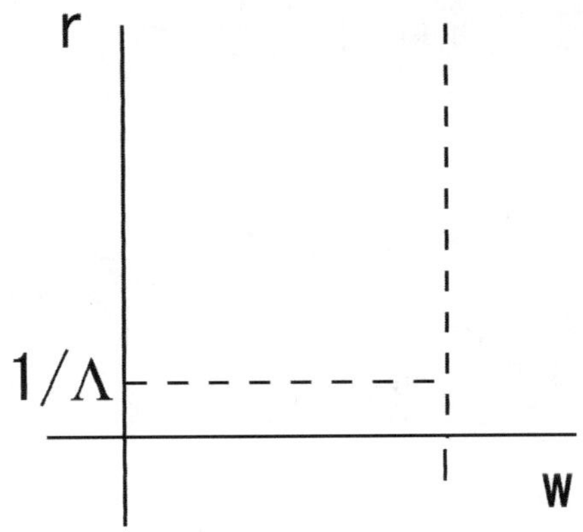

Figure 5. The integral region of (49).

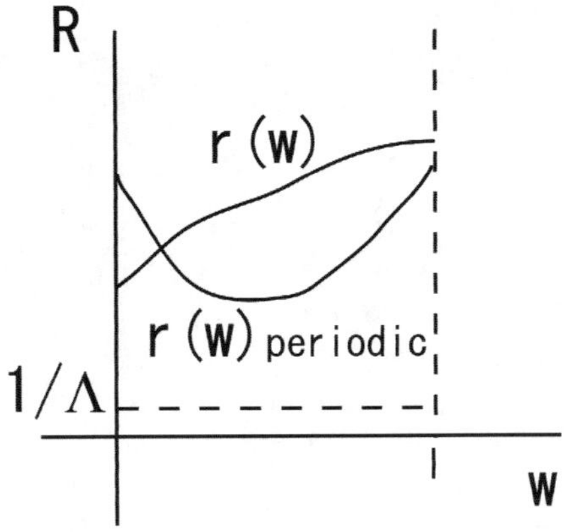

Figure 6. A general path $r(w)$ of (50) and a periodic path $r(w)$ of (51).

We regularize the above expression using the path-integral as

$$
E^{\mathcal{W}\ \prime}_{Cas}/(2L)^2 = \frac{1}{2^2\pi^3}(2\pi)\int_{\text{all paths } r(w)}
$$
$$
\prod_w \mathcal{D}r(w)\left[\int dw' P(\frac{1}{r(w')}) r(w')^{-4}\right]\exp\{-\mathcal{W}[r(w)]\} \quad , \tag{50}
$$

where the integral is over all paths $r(w)$ which are defined between $0 \leq w \leq l$ and whose value is above Λ^{-1}, as shown in Fig.6.

$\mathcal{W}[r(w)]$ is some damping functional explained in the next paragraph. The case $\mathcal{W}[r(w)] = 0$ corresponds to (49). The slightly-more-restrictve regularization is

$$
E^{\mathcal{W}}_{Cas}/(2L)^2 = \frac{1}{2^2\pi^3}(2\pi)\int_{\Lambda^{-1}}^{\infty} d\rho \int_{r(0)=r(l)=\rho}
$$
$$
\prod_w \mathcal{D}r(w)\left[\int dw' P(\frac{1}{r(w')}) r(w')^{-4}\right]\exp\{-\mathcal{W}[r(w)]\} \quad \geq 0 \quad , \tag{51}
$$

where the integral is over all *periodic* paths. Note that the above regularization keep the positive-definite property.

Hence the present regularization mainly defined by the choice of $\mathcal{W}[r(w)]$. In order to specify it, we introduce the following metric in (R,w)-space.

$$ds^2 = \frac{1}{dw^2}(dR^2 + \Omega^2 R^2 dw^2)^2 \quad . \tag{52}$$

This is the same as that in Sec.2.2. On a path $R = r(w)$, the induced metric and the length L is given as follows. As the damping functional $\mathcal{W}[r(w)]$, we take the length L.

$$ds^2 = dw^2(r'^2 + \Omega^2 r^2)^2 \quad , \quad r' \equiv \frac{dr}{dw} \quad ,$$
$$L = \int ds = \int (r'^2 + \Omega^2 r^2) dw \quad , \quad \mathcal{W}[r(w)] \equiv \frac{1}{2\alpha} L = \frac{1}{2\alpha}\int (r'^2 + \Omega^2 r^2) dw \quad . \tag{53}$$

The two parameters α and Ω are considered as regularization ones. The limit $\alpha \to \infty$ corresponds to (49).

Numerical calculation can evaluate $E^{\mathcal{W}}_{Cas}$ (51), and we expect the following form[24, 25].

$$\frac{E^{\mathcal{W}}_{Cas}}{(2L)^2} = \frac{a}{l^3}(1 - 3c\ln\ (l\Lambda)) \quad , \tag{54}$$

where a and c are some constants. a should be positive because of the positive-definiteness of (51). The present regularization result has, like the ordinary renormalizable ones such as the coupling in QED, the log-divergence. The divergence can be renormalized into the boundary parameter l. This means l *flows* according to the renormalization group.

$$l' = l(1 - 3c\ln(l\Lambda))^{-\frac{1}{3}} \quad , \quad \beta \equiv \frac{d\ln(l'/l)}{d\ln\Lambda} = c \quad , \quad |c| \ll 1 \quad , \tag{55}$$

where β is the renormalization group function, and we assume $|c| \ll 1$. The sign of c determines whether the length separation increases ($c > 0$) or decreases ($c < 0$) as the measurement resolution becomes finer (Λ increases). In terms of the usual terminology, attractive case corresponds to $c > 0$, and repulsive case to $c < 0$.

5.2. 5D Flat and Curved Case

Let us mention the space-time quantization (quantum gravity) in relation to the motivation of the present work. The space-time geometry is specified by the metric tensor field $g_{\mu\nu}(x)$ which appears in the definition of the line element $(ds^2)_{4D} = g_{\mu\nu}(x)dx^\mu dx^\nu$ $(\mu,\nu = 0,1,2,3)$. One of most important problems of the present theoretical physics is the clarification of the *quantum role* of the metric (gravitational) field $g_{\mu\nu}$. We already have a long (nearly half century) history of the quantum gravity since Feynman[20] and DeWitt [21] pioneered. About one decade ago, inspired by the development of the string theory and the D-brane theory, a fascinating model of unification of forces was proposed. It is a 5 dimensional(dim) model with AdS_5 geometry and is called "Randall-Sundrum model" or the "warped model"[22]. This is a representative of the extra dimensional models. An important purpose of the present work is to make this 5 dim model *legitimate* as the *quantum field theory*.

In the warped space-time , the geometry is described as

$$\text{Warped Metric (y-expression)} \quad ds^2 = e^{-2\omega|y|}\eta_{\mu\nu}dx^\mu dx^\nu + dy^2 \ , \ -l \le y \le l \ , \tag{56}$$

where $\{\mu,\nu = 0,1,2,3\}$, $(\eta_{\mu\nu}) = \text{diag}(-1,1,1,1)$. The coordinate y is called the extra coordinate. The parameter ω is the 5 dim (bulk) scalar curvature. l is the size parameter of the extra coordinate. We respect the periodicity: $y \to y+2l$, and Z_2-parity: $y \leftrightarrow -y$. Instead of y, another coordinate z is also used. [8]

$$\text{Warped Metric (z-expression)} \quad ds^2 = \frac{1}{\omega^2 z^2}(\eta_{\mu\nu}dx^\mu dx^\nu + dz^2) = G_{MN}dX^M dX^N \ ,$$
$$|z| = \frac{1}{\omega}e^{\omega|y|} \ , \quad \frac{1}{\omega} < |z| < \frac{1}{T} \ , \quad T \equiv \omega e^{-\omega l} \ ,$$
$$R_{MN} = 4\omega^2 G_{MN} \ , \quad R = 20\omega^2 > 0 \ , \quad \sqrt{-G} = \sqrt{-\det G_{MN}} = \frac{1}{(\omega|z|)^5} \ , \qquad (57)$$

where $(X^M) \equiv (x^\mu, z)$, $\{M, N = 0,1,2,3,5\}$. [9] The flat (5D Minkowski) limit is obtained by $\omega \to 0$ in the y-expression (56).

$$\text{Flat Metric} \quad ds^2 = \eta_{\mu\nu}dx^\mu dx^\nu + dy^2 \ , \quad (X^M) = (x^\mu, y) \ , \quad -l \le y \le l \ , \qquad (58)$$

Traditional calculation[23, 24, 25] gives the Λ^5-divergent result for Casimir energy, on the above geometries, of 5D models. In the calculation, Casimir energy is expressed as the 5D space-momentum integral ($\int d^4p_E dy$ or $\int d^4p_E dz$) of some energy (density) function $F(\tilde{p}, y)$ or $F(\tilde{p}, z)$. (See Appendix for detail.) In ref.[24, 25], we claim the Λ^5-divergence comes from this 'naive' integration measure and should be replaced by some proper measure, based on close numerical calculation using some trial integration measures. Finally, Casimir energy of the free fields (electromagnetism, free scalar theory) is *proposed* to be replaced by the following *path-integral.*

$$\text{For Flat Geometry :}$$
$$-\mathcal{E}_{Cas}(l,\Lambda) = \int_{1/\Lambda}^{l} d\rho \int_{r(0)=r(l)=\rho} \prod_{a,y} \mathcal{D}x^a(y) \left[\int dy' F_1(\frac{1}{r(y')}, y')\right]$$
$$\times \exp\left[-\frac{1}{2\alpha'}\int_0^l \sqrt{r'^2+1}\, r^3 dy\right] , \ r' = \frac{dr}{dy},$$
$$\text{For Warped Geometry :}$$
$$-\mathcal{E}_{Cas}(\omega,T,\Lambda) = \int_{1/\Lambda}^{1/\mu} d\rho \int_{r(1/\omega)=r(1/T)=\rho} \prod_{a,z} \mathcal{D}x^a(z) \left[\int dz' F_2(\frac{1}{r(z')}, z')\right]$$
$$\times \exp\left[-\frac{1}{2\alpha'}\int_{1/\omega}^{1/T} \frac{1}{\omega^4 z^4}\sqrt{r'^2+1}\, r^3 dz\right] \ , \quad r' = \frac{dr}{dz} \ , \qquad (59)$$

where $r = \sqrt{\sum_{a=1}^4 (x^a)^2}$. [10] ($\{x^a | a = 1,2,3,4\}$ is the *Euclideanized* coordinates of $\{x^\mu | \ \mu = 0,1,2,3\}$, $x^0 = ix^4$.) F_1 and F_2 are some energy density functions and will appear in Appendix ((A.2) and (A.4)). Λ is the UV-cutoff parameter, $\mu \equiv \Lambda T/\omega$ is the IR-cutoff one and l is the

[8] z is defined by y as

$$z = \begin{cases} \frac{1}{\omega}e^{\omega y} & y > 0 \\ 0 & y = 0 \\ -\frac{1}{\omega}e^{-\omega y} & y < 0 \end{cases}$$

[9] T is *not* a temperature parameter but a IR parameter like l ($T = \omega e^{-\omega l}$). The temperature appears as β^{-1}. See eq.(3).

[10] The case $\alpha' \to \infty$ in (59) is essentially the traditional definition of Casimir energy.

periodicity(IR) one. Note that the above expressions of $-\mathcal{E}_{Cas}$ are positive-definite. The above path-integrals are over all paths of 4 dim *hypersurfaces* defined by

$$\text{Flat Geometry:} \quad \sqrt{\sum_{a=1}^{4}(x^a)^2} = r(y) \quad , \quad -l \leq y \leq l \quad ,$$
$$\text{Warped Geometry:} \quad \sqrt{\sum_{a=1}^{4}(x^a)^2} = r(z) \quad , \quad \frac{1}{\omega} \leq |z| \leq \frac{1}{T} \quad . \tag{60}$$

See Fig.7 for the case of the N+1 dim space. This is a *closed string* configurartion. The *area* (4D

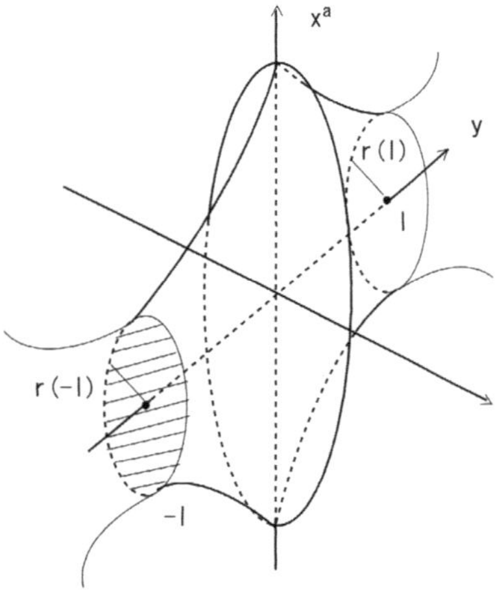

Figure 7. N(=2) dim hypersurface in N+1 dim (Euclidean flat) space $(x^1, x^2, \cdots, x^N, y)$. Sphere S^{N-1}(circles in the figure) at y has the radius $r(y)$.

volume) plays the role of *Hamiltonian* of the quantum statistical system $\{x^a\}$. F_i comes from the *matter-field* quantization and plays a role of the *energy 'operator'* in the path-integral over the 4D hyper-surface $r(y)$ or $r(z)$. The *string (surface) tension* parameter $1/2\alpha'$ is introduced. The new point, compared with the 5D Casimir energy calculation so far[23], is the introduction of the 'minimal area' factor $\exp(-\frac{1}{2\alpha'}\text{Area}) = \exp(-\frac{1}{2\alpha'}\int\sqrt{\det(g_{ab})}d^4x)$ where g_{ab} is the *induced* metric on the hyper-surface (60). We have shown, in this paper, the above-type *path-integral* very naturally appears in many quantum-statistical systems when we view them *geometrically.* The proposed quantities (59) are shown to be valid.

The proposed expressions (59) of 5D Casimir energy can be evaluated numerically. We confirmed, not using the path-integral but using some effective approach (weight-function method), that they are given by as follows.

$$\text{Flat} \ : \ \frac{\mathcal{E}_{Cas}(l,\Lambda)}{\Lambda l} = -\frac{a}{l^4}(1 - 4c\ln(l\Lambda)) \ , \quad a \sim 2.5 \quad , \quad c > 0 \quad , \quad c \sim O(10^{-3}) \quad ,$$
$$\text{Warped} \ : \ \frac{\mathcal{E}_{Cas}(\omega,T,\Lambda)}{\Lambda T^{-1}} = -a\omega^4(1 - 4c_1\ln(\frac{\Lambda}{\omega}) - 4c_2\ln(\frac{\Lambda}{T})) \ ,$$
$$a \sim 1.2 \ , \ c_1 \sim -0.11 < 0 \ , \ c_2 \sim 0.10 > 0 \ . \tag{61}$$

The boundary parameters *flow* as

$$\text{Flat} \ : \ \frac{\mathcal{E}_{Cas}(l,\Lambda)}{\Lambda l} = -\frac{a}{l'^4} \quad , \quad \beta = \frac{\partial}{\partial(\ln\Lambda)}\ln\frac{l'}{l} = c > 0 \quad ,$$
$$\text{Warped} \ : \ \frac{\mathcal{E}_{Cas}(\omega,T,\Lambda)}{\Lambda T^{-1}} = -a\omega'^4 \quad , \quad \beta = \frac{\partial}{\partial(\ln\Lambda)}\ln\frac{\omega'}{\omega} = -c_1 - c_2 \quad . \tag{62}$$

$\beta = c > 0$ in the flat case means the size of the extra world (periodicity l) shrinks as the measurement resolution becomes coarse. This image fits with the compactification of the extra axis in the higher-dimensional unified models.

6. Visco-Elastic System

The present approach gives a new method for the study of the visco-elastic system. Let us explain it using the harmonic oscillator with friction.

$$m\ddot{x} = -kx - \eta\dot{x} \ , \quad k: \text{ spring constant}, \quad \eta: \text{ viscosity} \quad . \tag{63}$$

From this we obtain the "energy" at $t = t_0$.

$$\mathcal{E} = \left(\frac{1}{2}\dot{x}^2 + \frac{{\omega_1}^2}{2}x^2\right)|_{t_0} = \frac{1}{2}\dot{x}^2 + \frac{{\omega_1}^2}{2}x^2 + \eta' \int_{t_0}^{t} \left(\frac{dx(\tilde{t})}{d\tilde{t}}\right)^2 d\tilde{t} \ , \quad {\omega_1}^2 = \frac{k}{m}, \ \eta' = \frac{\eta}{m}. \tag{64}$$

This quantity is conserved (independent of t). The hysteresis term in the above expression represents the energy from the friction force. We can read the line elements in (X, t) space for this system.

$$\begin{aligned} &\text{'Dirac' type}: \quad ds^2 = dX^2 + dt^2 \left({\omega_1}^2 X^2 + 2\eta' \int_{t_0}^{t} (dX)^2 \frac{1}{d\tilde{t}}\right) \quad , \\ &\text{Standard type}: \quad ds^2 = \frac{1}{dt^2}\left\{dX^2 + dt^2 \left({\omega_1}^2 X^2 + 2\eta' \int_{t_0}^{t} dX^2 \frac{1}{d\tilde{t}}\right)\right\}^2 \end{aligned} \tag{65}$$

On a path $X = x(t)$, with the standard type, the *length* $L[x(t)]$ and the free energy F is given by $(t_0 = 0)$

$$\begin{aligned} L[x(t)] &= \int ds = \int_0^{\beta} dt\{\dot{x}^2 + {\omega_1}^2 x^2 + 2\eta' \int_0^t (\frac{dx(\tilde{t})}{d\tilde{t}})^2 d\tilde{t}\} \quad , \\ \mathrm{e}^{-\beta F(l,\beta)} &= \int_{-l}^{l} d\rho \int_{x(0)=\rho, x(\beta)=\rho} \mathcal{D}x(t) \mathrm{e}^{-\frac{1}{2}L[x(t)]} \quad , \end{aligned} \tag{66}$$

where $2l$ is the periodicity, with which we impose X periodic.

$$X \to X + 2l \quad . \tag{67}$$

The energy E, the entropy S and the force f are given by

$$\begin{aligned} &\text{Energy} \quad E(l,\beta) = < \frac{L}{2} > = \int_{-l}^{l} d\rho \int \mathcal{D}x(t) \frac{L[x(t)]}{2} \exp\{-\frac{1}{2}L[x(t)]\} \quad , \\ &\text{Entropy} \quad S(l,\beta) = k_B \beta (E(l,\beta) - F(l,\beta)) \quad , \quad \text{Force} \quad f(l,\beta) = -\frac{\partial E(l,\beta)}{\partial l} \quad , \end{aligned} \tag{68}$$

where k_B is Boltzmann's constant.

The simple model (63) can be generalized as

$$m\ddot{x} = -\frac{\partial V(x)}{\partial x} - \eta_1 \dot{x} W_1(x) - \eta_2 \dot{x}^2 W_2(x) \quad , \tag{69}$$

where we assume the visco-elastic system is in the relatively slow motion. This assumption makes the velocity($\dot{x}$)-expansion of (69) valid. $V(x), W_1(x), W_2(x)$ are general functions. There are many choices:

$$\begin{aligned} V(x) &= \frac{1}{2}kx^2 \text{ (spring)},\ gx \text{ (rain-drop)},\ \cdots \\ W_1(x) &= 1 \quad , \quad x^2-1 \text{ (van der Pol eq.)},\ \cdots \\ W_2(x) &= 1 \quad , \quad x \quad , \quad x^2 \quad , \quad x^3 \quad , \quad \cdots \end{aligned} \tag{70}$$

The line elements in (X,t) space is given by

$$\begin{aligned} &\text{'Dirac' type :} \\ ds^2 &= m\ dX^2 + dt^2\left\{2V(X) + 2\eta_1\int_{t_0}^{t}(dX)^2\frac{1}{d\tilde{t}}W_1(X) + 2\eta_2\int_{t_0}^{t}(dX)^3\frac{1}{(d\tilde{t})^2}W_2(X)\right\}, \\ &\text{Standard type :} \\ ds^2 &= \frac{1}{dt^2}\left[m\ dX^2 + dt^2\left\{2V(X) + 2\eta_1\int_{t_0}^{t}(dX)^2\frac{1}{d\tilde{t}}W_1(X)\right.\right. \\ &\qquad \left.\left. +2\eta_2\int_{t_0}^{t}(dX)^3\frac{1}{(d\tilde{t})^2}W_2(X)\right\}\right]^2 . \end{aligned} \tag{71}$$

The above formulas are valid for the spacially-circling 1-dim visco-elastic fluid (Fig.8). The

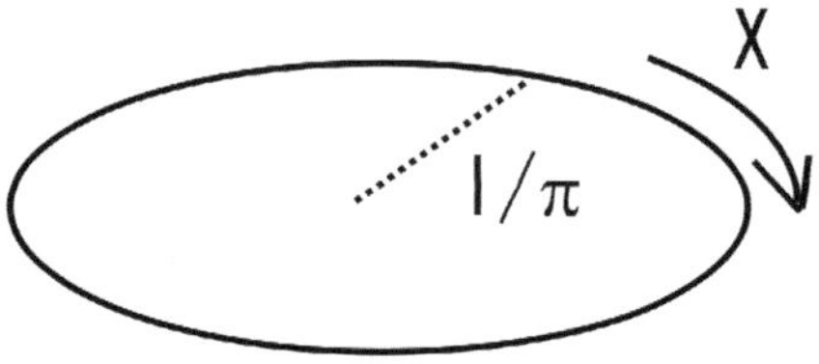

Figure 8. One dimensional, circling, visco-elastic flow. The radius l/π is sufficiently large.

global (statistical) physical quantities can be expressed by the path-integral in the same way as eq.(68).

For the spacially 3 dim visco-elastic system, we treat Navier-Stokes equation (1). We obtain the following relation.

$$\frac{D}{Dt}\left(\frac{\rho}{2}\vec{v}^2 + P - \rho\vec{x}\cdot\vec{g}\right) = \frac{\partial P}{\partial t} + \eta v^i\Delta v^i \quad . \tag{72}$$

From this, we obtain, for the case $\frac{\partial P}{\partial t} = 0$,

$$\mathcal{E} = \left(\frac{\rho}{2}\vec{v}^2 + P - \rho\vec{x}\cdot\vec{g}\right)|_{t_0} = \frac{\rho}{2}\vec{v}^2 + P - \rho\vec{x}\cdot\vec{g} - \eta\int_{t_0}^{t}v^i\Delta v^i d\tilde{t} \quad . \tag{73}$$

We can read, from the above result, the following line elements.

$$\begin{aligned} &\text{Dirac type :}\quad ds^2 = \rho\ dX^{i\ 2} + dt^2\left\{2P(X) - 2\rho X^i g^i - 2\eta\int_{t_0}^{t}dX^i\frac{1}{d\tilde{t}}\Delta(dX^i\frac{1}{d\tilde{t}})\ d\tilde{t}\right\}, \\ &\text{Standard type :} \\ &ds^2 = \frac{1}{dt^2}\left[\rho\ dX^{i\ 2} + dt^2\left\{2P(X) - 2\rho X^i g^i - 2\eta\int_{t_0}^{t}dX^i\frac{1}{d\tilde{t}}\Delta(dX^i\frac{1}{d\tilde{t}})\ d\tilde{t}\right\}\right]^2 . \end{aligned} \tag{74}$$

On a path $X^i = x^i(t)$, with the standard type, the *length* $L[\vec{x}(t)]$ and the free energy F are given by $(t_0 = 0)$

$$L[\vec{x}(t)] = \int ds = \int_0^\beta dt \left\{ \rho(\dot{x}^i)^2 + 2P(\vec{x}) - 2\rho\vec{x}\cdot\vec{g} - 2\eta\int_0^t \dot{x}^i\Delta\dot{x}^i\, d\tilde{t} \right\} \quad ,$$

$$e^{-\beta F(l,\beta)} = \int_{-l}^{l} d\rho_1 \int_{-l}^{l} d\rho_2 \int_{-l}^{l} d\rho_3 \int_{x^i(0)=x^i(\beta)=\rho^i} \mathcal{D}x^i(t) e^{-\frac{1}{2}L[\vec{x}(t)]} \quad , \qquad (75)$$

where $2l$ is the periodicity, with which we impose X^i periodic.

$$X^i \to X^i + 2l \qquad (i = 1, 2, 3) \quad . \qquad (76)$$

7. Discussion and Conclusion

Inspired by the recent new views on the quantum gravity, we have presented a geometrical approach to the general quantum statistical system. The idea lies in the introduction of the metric in the space of Feynman's path-integral. The length or the area gives the system Hamiltonian. In the application to 5D Casimir energy, the inverse temperature (or proper time) axis is played by the extra (5th) one. In the 4D case (ordinary Casimir energy), the regularization axis plays the role. Casimir force is explained from the *renormalization group flow*. The attractive force or the repulsive one corresponds to the positive β-function or the negative one. This geometrical approach also gives a new method to examine the dissipative system caused by friction. Taking simple visco-elastic models, we elaborate on how to choose the metric and present the path-integral expressions of the statistical quantities such as energy and entropy.

Appendix A. Traditional Definition of Casimir Energy in 5D Theories

The traditional definition of Casimir Energy of the 5D electromagnetic field theory is, for the flat case (58),

$$e^{-l^4 E_{Cas}} = \int \mathcal{D}A \exp\left[i\int d^4x dy (\mathcal{L}_{EM}^{5D} + \mathcal{L}_{gauge}) \right]\Big|_{\text{Euclid}} \quad ,$$

$$\mathcal{L}_{EM}^{5D}[A_M(X)] = -\frac{1}{4}F_{MN}F^{MN} \ , \ F_{MN} = \partial_M A_N - \partial_N A_M \ , \ \mathcal{L}_{gauge}[A_M(X)] = -\frac{1}{2}(\partial_M A^M)^2 \quad (A.1)$$

The expression of E_{Cas} defined above, is given by

$$\text{For Flat Geometry (5 dim elctromagnetism)} :$$

$$E_{Cas}(l) = \int_{\tilde{p}\leq\Lambda} \frac{d^4p}{(2\pi)^4} \int_0^l dy (F_f^-(\tilde{p}, y) + 4F_f^+(\tilde{p}, y)) \quad ,$$

$$F_f^{\mp}(\tilde{p}, y) = -\int_{\tilde{p}}^{\infty} d\tilde{k} \frac{\mp\cosh\tilde{k}(2y - l) + \cosh\tilde{k}l}{2\sinh(\tilde{k}l)} \quad . \qquad (A.2)$$

The plus-minus symbol, $\mp$, indicates the contribution from Z_2-parity odd (-) and even (+) components. $\tilde{p}$ is the maginitude of 4D momentum $(p_a) = (p_1, p_2, p_3, p_4)$. The coincidence with the previous result[23] was confirmed[24]. As for the warped case (57), the traditional definition, for the 5D free scalar theory, is given by

$$e^{-T^{-4}E_{Cas}} = \int \mathcal{D}\Phi \ \exp\left[i\int d^5X \sqrt{-G}\mathcal{L}_s^{5D} \right]\Big|_{\text{Euclid}}$$

$$= \int \mathcal{D}\Phi(X) \exp\left[\int d^4x dz \frac{1}{(\omega z)^5}\frac{1}{2}\Phi\{\omega^2 z^2 \partial_a\partial^a\Phi + (\omega z)^5 \hat{L}_z\Phi\} \right] \quad ,$$

$$\mathcal{L}_s^{5D}[\Phi(X);X] = -\frac{1}{2}\nabla^M\Phi\nabla_M\Phi - \frac{1}{2}m^2\Phi^2 \quad ,$$
$$\frac{1}{\omega} < |z| < \frac{1}{T} \quad , \quad \hat{L}_z = \frac{d}{dz}\frac{1}{(\omega z)^3}\frac{d}{dz} - \frac{m^2}{(\omega z)^5} \quad , \quad (m^2 = -4\omega^2) \quad . \tag{A.3}$$

where $\hat{L}_z$ is the kinetic operator in the extra space (Bessel differential operator). Casimir energy E_{Cas} defined in (A.3) is explicitly given by

For Warped Geometry (5 dim Free Scalar, $m^2 = -4\omega^2$):

$$-E^{\mp}_{Cas}(\omega,T) = \int \frac{d^4p_E}{(2\pi)^4}\bigg|_{\tilde{p}\leq\Lambda} \int_{1/\omega}^{1/T} dz\ F^{\mp}_w(\tilde{p},z) \quad , \quad F^{\mp}_w(\tilde{p},z) = \frac{1}{(\omega z)^3}\int_{\tilde{p}^2}^{\infty}\{G^{\mp}_k(z,z)\}dk^2 \quad ,$$
$$G^{\mp}_p(z,z') = \mp\frac{\omega^3}{2}z^2z'^2\frac{\{\mathbf{I}_0(\frac{\tilde{p}}{\omega})\mathbf{K}_0(\tilde{p}z) \mp \mathbf{K}_0(\frac{\tilde{p}}{\omega})\mathbf{I}_0(\tilde{p}z)\}\{\mathbf{I}_0(\frac{\tilde{p}}{T})\mathbf{K}_0(\tilde{p}z') \mp \mathbf{K}_0(\frac{\tilde{p}}{T})\mathbf{I}_0(\tilde{p}z')\}}{\mathbf{I}_0(\frac{\tilde{p}}{T})\mathbf{K}_0(\frac{\tilde{p}}{\omega}) - \mathbf{K}_0(\frac{\tilde{p}}{T})\mathbf{I}_0(\frac{\tilde{p}}{\omega})},$$
$$(\hat{L}_z - p^2 s(z))G^{\mp}_p(z,z') = \begin{cases} \epsilon(z)\epsilon(z')\hat{\delta}(|z|-|z'|) & \text{for P}=-1 \\ \hat{\delta}(|z|-|z'|) & \text{for P=1} \end{cases} \quad , \quad s(z) = \frac{1}{(\omega z)^3} \tag{A.4}$$

where $\mathbf{I}_0$ and $\mathbf{K}_0$ are the modified Bessel functions of 0-th order.

Casimir energy defined above, which has been traditionally calculated, gives Λ^5-divergence. The integral $\int \frac{d^4p_E}{(2\pi)^4}dz$ ($\int \frac{d^4p}{(2\pi)^4}dy$) appearing in eq.(A.4) ((A.2)) corresponds to the summation over all positions in 5 dim bulk space $\int d^4xdz$ ($\int d^4xdy$). The above expression says E_{Cas} is the total sum of $F(r^{-1},z)$ ($F(r^{-1},y)$) over the bulk space positions. We notice here the Λ^5 divergence comes from the fact that we have overlooked some proper *integration measure*. The summation, or the *averaging* procedure (of F) should be properly defined at this stage. In the present standpoint we regard the *coordinate* system (x^a,z) $((x^a,y))$ as the *quantum statistical* system and consider that the coordinate x^a is the *quantum mechanical* variable with the extra one z (y) as *Euclidean time*. The traditional treatment (simple summation over the set of positions) should be corrected by the present quantum (geometric) approach. We have proposed it should be done by the *path-integral* over all hypersurfaces in the bulk space (x^a,z) $((x^a,y))$, as described in Sec.5. Hence the right expression of Casimir energy is given by (59).

Acknowledgments

Parts of the content of this work have been already presented at the international conference on "Particle Physics, Astrophysics and Quantum Field Theory"(08.11.27-29, Nanyang Executive Centre, Singapore)[26], First Mediterranean Conference on Classical and Quantum Gravity (09.9.14-18, Kolymbari, Crete, Greece)[27], Int. Workshop on "Strong Coupling Gauge Theories in LHC Era"(09.12.8-11, Nagoya Univ., Nagoya, Japan)[28], and IPMU Workshop on 'Condensed Matter Physics Meets High Energy Physics' (10.2.8-12, IPMU, Univ. of Tokyo, Kashiwa, Japan). The author thanks the audience for useful comments and encouragement on the occasions.

References

[1] M. B. Green, J. H. Schwartz and E. Witten, *Superstring theory, Vol.I and II*, Cambridge Univ. Press, c1987, Cambridge
J. Polchinski, *STRING THEORY, Vol.I and II*, Cambridge Univ. Press, c1998, Cambridge

[2] J.M. Maldacena, *Adv.Theor.Math.Phys.***2**(1998)231 [*Int. J. Theor. Phys.***38**(1999)1113](*Preprint* arXiv:hep-th/9711200)

[3] S. S. Gubser, I. R. Klebanov and A. M. Polyakov, *Phys. Lett.* **B428**(1998)105(*Preprint* arXiv:hep-th/9802109)

[4] E. Witten, *Adv. Theor. Math. Phys.***2**(1998)253 (*Preprint* arXiv:hep-th/9802150)

[5] M. Natsuume, "String theory and quark-gluon plasma" (*Preprint* arXiv:hep-ph/0701201)
[6] D.T. Son, *Ann.Rev.Nucl.Part.Sci.***57**(2007)95 (*Preprint* arXiv:0704.0240[hep-th])
[7] D. Mateos, *Class. Quantum Grav.* **24**(2007)S713 (*Preprint* arXiv:0709.1523[hep-th])
[8] S.S. Gubser, *Phys. Rev.* **D78**(2008)065034
[9] S.A. Hartnoll, C.P. Herzog and G.T. Horowitz, *Phys. Rev. Lett.* **101**(2008)031601
[10] S.S. Gubser and S.S. Pufu, *J. High Energy Phys.***0811**(2008)033
[11] S.A. Hartnoll, C.P. Herzog and G.T. Horowitz, *J. High Energy Phys.***0812**(2008)015
[12] T. Sakai and S. Sugimoto, *Prog.Theor.Phys.***113**(2005)843 (*Preprint* arXiv:hep-th/0412141)
[13] T. Sakai and S. Sugimoto, *Prog.Theor.Phys.***114**(2005)1083 (*Preprint* arXiv:hep-th/0507073)
[14] P. Hořava, "Membranes at Quantum Criticality", *J. High Energy Phys.***0903**(2009)020 (*Preprint* arXiv:hep-th/0812.4287)
[15] P. Hořava, "Quantum Gravity at a Lifshitz Point", *Phys. Rev.* **D79**(2009)084008 (*Preprint* arXiv:hep-th/0901.3775)
[16] E. Verlinde, "On the Origin of Gravity and the Laws of Newton" (*Preprint* arXiv:1001.0785[hep-th])
[17] A.M. Polyakov, *Phys. Lett.* **103B**(1981),207
[18] R.P. Feynman, "Statistical Mechanics", W.A.Benjamin,Inc., Massachusetts, 1972
[19] W. F. Schelter, **Maxima 5.19 Manual**, GNU project, c2009
[20] R. P. Feynman, *Acta Phys. Polonica* **24**, 697(1963)
[21] B. S. DeWitt, *Phys. Rev.***162**, 1195, 1239(1967)
[22] L. Randall and R. Sundrum, *Phys. Rev. Lett.* **83**(1999)3370,4690
[23] T. Appelquist and A. Chodos, *Phys. Rev.* **D28**(1983)772
T. Appelquist and A. Chodos, *Phys. Rev. Lett.* **50**(1983)141
[24] S. Ichinose, *Prog.Theor.Phys.***121**(2009)727 (*Preprint* ArXiv:0801.3064v8[hep-th])
[25] S. Ichinose, "Casimir Energy of 5D Warped System and Sphere Lattice Regularization", *Preprint* ArXiv:0812.1263[hep-th], US-08-03, 61 pages.
[26] S. Ichinose, "Casimir Energy of AdS5 Electromagnetism and Cosmological Constant Problem", *Int.Jour.Mod.Phys.***24A**(2009)3620, Proc. of Int. Conf. on Particle Physics, Astrophysics and Quantum Field Theory: 75 Years since Solvay (Nov.27-29, 2008, Nanyang Executive Centre, Singapore), *Preprint* arXiv:0903.4971
[27] S. Ichinose, "Casimir Energy of the Universe and New Regularization of Higher Dimensional Quantum Field Theories", First Mediterranean Conference on Classical and Quantum Gravity (09.9.14-18, Kolymbari, Crete, Greece), *Jour.Phys.:Conf.Ser.***222**(2010) 012048 (*Preprint* ArXiv:1001.0222[hep-th])
[28] S. Ichinose, "New Regularization in Extra Dimensional Model and Renormalization Group Flow of the Cosmological Constant", Int. Workshop on 'Strong Coupling Gauge Theories in LHC Era'(09.12.8-11, Nagoya Univ., Nagoya, Japan). to appear in the proceedings. *Preprint* ArXiv:1003.5041[hep-th]

MEMS-in-TEM for Nano Tribology

Hiroyuki Fujita, **Tadashi Ishida, Takaaki Sato, Shinsuke Nabeya**

Institute of Industrial Science, The University of Tokyo, 4-6-1 Komaba, Meguro-ku, Tokyo, 153-8505, Japan

fujita@iis.u-tokyo.ac.jp

Abstract. We have developed a MEMS-in-TEM experimental system that enables the mechanical testing of nano junctions by MEMS microactuators in TEM (transmission electron microscope) for *in situ* atomic level observation of dynamic deformation. A nano junction between two sharp opposing tips was formed by bringing them into contact. Applying either tensile or shearing stress, we observed the elongation and changes in shape of the junction at real-time rate while applied force was measured. The experimental system and results of tensile and shear tests are described.

1. Introduction

We have intensively investigated MEMS (micro electro mechanical system) design, fabrication and its application to nano and bio technologies [1]. Our nano scientific research using MEMS devices includes the mechanical testing of nano junctions in TEM (transmission electron microscope) for *in situ* atomic level observation of dynamic deformation [2]. Test structures in the MEMS are as small as ten nanometers. MEMS electrostatic actuators can provide higher precision at the sub-nm scale, more stable loading control over long time (hours, even days), and better temperature stability than conventional piezoelectric actuators that are used commonly in atomic force microscopes. Furthermore, the integration of thermal or strain sensors or heaters enables multi-functional MEMS devices that can fit in the narrow space of a TEM specimen holder. The tensile testing device has opposing sharp tips; one tip is fixed and another is actuated toward it. By bringing them together, a nano junction is formed. The junction is stretched and separated by retracting the movable tip.

This time, we fabricated a two-DOF (degree-of-freedom) device that had another actuator to apply shear stress to the junction for tribology measurement. The shape and dimension of nano junctions as well as the displacement of the movable tip were observed during the shear testing. We prepared tips of various materials that showed very different behaviors. The applied force can be calculated from the spring constant of the support and the difference in displacement between the cases with and without a junction.

2. MEMS-in-TEM experimental system

As shown in figure 1, a MEMS device driven inside a TEM specimen chamber constitutes a MEMS-in-TEM system. The MEMS device was placed in a special TEM holder with nine feed-throughs and inserted into the specimen chamber of the TEM (Hitachi, HF-2000UHV), with a vacuum level of less than 1×10^{-7} Pa. The feed-throughs of the holder enabled the actuation of the MEMS device and electrical current measurements between tips during *in situ* TEM observation. The electron beam accelerated at 200 kV irradiated and passed through the gap of the opposing tips. The sample thickness should be less than a few hundreds of nanometers because the electron beam must penetrate the sample for observation with TEM [3]. High-resolution images were projected on a screen with 0.2 nm maximum spatial resolution. The electric current density in the observation area of the TEM electron beam was 6.2×10^4 A m^{-2}. The electron beam of this magnitude causes temperature rise calculated to be less than 1 K [2, 3]. The images were captured by an integrated charge capacitive detection camera (SC200, Gatan, Inc.). High-resolution images could be obtained without scanning the electron beam and therefore allowed real-time observation. A temporal resolution of 1/30 s was achieved with a video recording system.

Our experiments were performed between sharp probe tips of less than 100 nm in radius of curvature. Such sharp tips were formed at the ends of micromachined structures. In order to ensure good alignment between sharp edges for the TEM observation, we designed the device with a symmetrical actuator and supports having proper rigidity. A schematic view of the device is shown in figure 1. The device has a movable electrode, a fixed electrode, a pair of driving electrodes and opposing tips. All the components were integrated within the limited volume, 4 mm x 5.4 mm x 0.7 mm, because the device must be inserted into the objective lens of the TEM. Opposing tips consisted of a movable tip and a fixed tip. Those tips were located at the center of the movable and fixed electrode, respectively. The electrostatic microactuator had two electrodes to drive the movable tip towards the fixed tip. Basic tip material was silicon. They could easily be covered with a thin layer of other materials such as Au, Ag, or DLC [4-6]. If required, the electrical current measurement between the opposing tips was possible when bias voltage was applied between them.

The whole device was micromachined from a silicon-on-insulator (SOI) substrate. Therefore, the thermal expansion occurs uniformly when environmental temperature may change. We also characterized the precision of the electrostatic actuator. The parabolic relationship between driving voltage and displacement was obtained from TEM direct observation. No hysteresis appeared. The displacement around 300 nm, where tips were almost brought into contact, could be approximated with a straight line; its slope was 6.2 nm V^{-1}. The driving voltage source had the precision of 0.0001 V (E5263A, Agilent Technologies). Hence, the tip displacement could be controlled within less than sub-nm order accuracy. This driving characteristic was found to be persistent over several experimental runs.

The displacement became smaller when the same characteristic was measured after the formation of a nano junction. The applied force can be determined by multiplying the spring constant of the support and the difference in displacement between the cases with and without a junction.

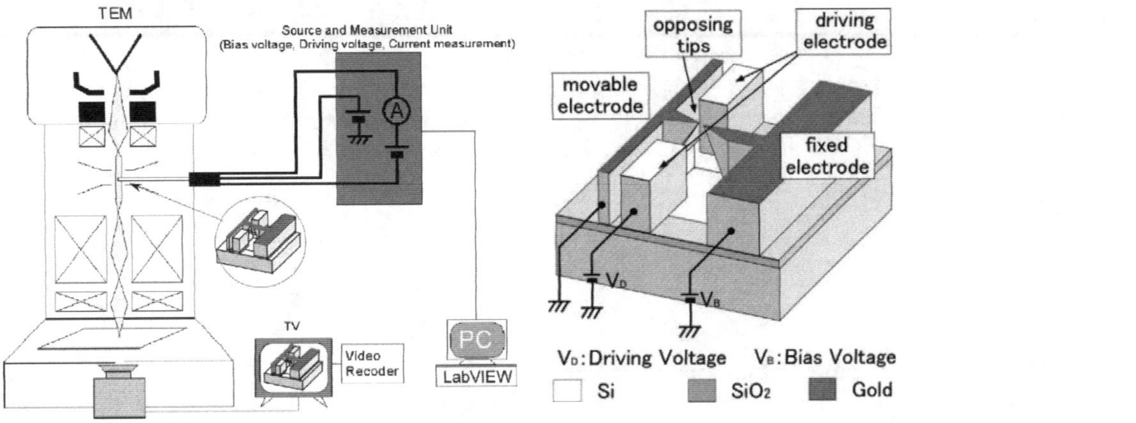

Figure 1 The schematic illustration of TEM observation system (left). A MEMS device having movable opposing tips (right) was inserted into the TEM specimen chamber. TEM real-time images were obtained with the TV camera.

3. Tensile test of nano junction by MEMS-in-TEM system

The tensile test of a gold nano junction [4, 7] is described to show the operational procedure of nano junction formation between tips and its mechanical testing under TEM observation. Here the tips in the MEMS device were covered with chromium/gold stacking layers of 5 nm/200 nm in thickness.

The procedure of our experiment, approach-contact-retraction-fracture process, is as follows:

a) The movable tip approaches the fixed tip by increasing the driving voltage.
b) A gold nano-bridge appears between tips when the gap distance is less than 0.5 nm at 1 V bias voltage applied between tips. The bridge enlarges by the atomic rearrangement.
c) The gold nano- junction was retracted. The diameter of the junction is thinned.
d) The junction is finally fractured. Corners on tips are rounded by the atomic rearrangement.

This process was repeated many times. The surface geometry was changed during this process resulting in the change of the contact surface [4]. Retracted shapes of gold nano junctions were different according to the elongation directions (figure 3) [8]. The crystalline orientation was decided from the TEM lattice image.

The gold nano junction was retracted keeping the similar figures when the elongation direction was [011]. The waist of the gold nano junction became thin from 8.8 nm to 2 nm in diameter (figures 3 a-c). The gold nano junction was transformed into the wireshape when the elongation direction was [0$\bar{1}$1]. The wire diameter was 1.8 nm during the 2.8 nm retraction along the length direction (figure 2 b-1). These gold nano junctions were broken spontaneously in about 10 seconds while the tip was kept at the same position. After breakage the tip became rounded and its final radius curvature was 3 nm (figures 3 d). This fracture was caused due to the atomic rearrangement of gold atoms.

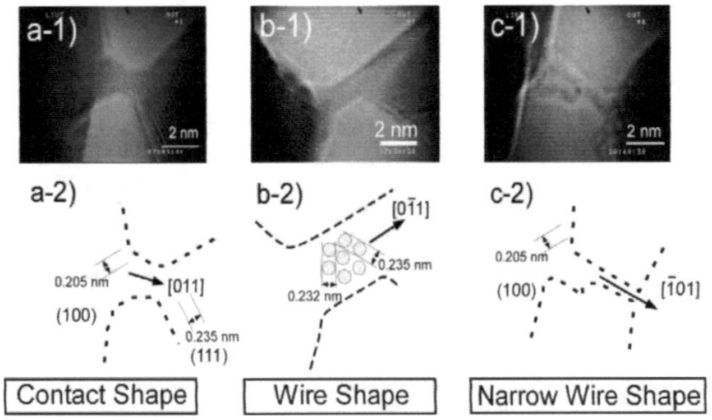

Figure 2 The identification method of the crystallographic facet. TEM images (a1, b1, c1) were schematically illustrated like in the images (a2, b2, c2). The elongation directions were [011], [0$\bar{1}$1], [$\bar{1}$01], respectively.

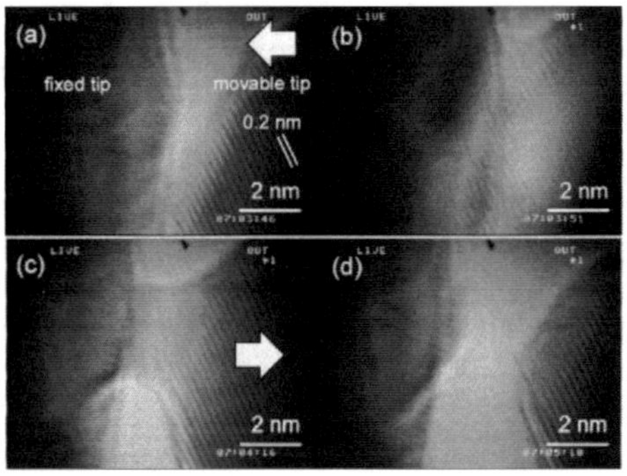

Figure 3 The gold nano junction elongation. When the direction of the gold nano junction elongation was [011] at the junction, the junction was thinned, holding the hourglass shape, and finally broken. The tips were rounded by the atom rearrangement.

4. MEMS-in-TEM for nano tribology experiment

4.1. MEMS device

As the extension of tensile-testing devices, we fabricated a two-DOF (degree-of-freedom) device that had another actuator to apply shear stress to the junction for tribology measurement. Its schematic illustration is shown in figure 4. One tip can move in longitudinal direction by the same actuator as in the tensile-test device. Another tip is driven in lateral direction by the additional actuator (electrode 3 in figure 4) placed on the side of the tip support. We prepared tips of bare Si and those coated with Au, Ag, and DLC (diamond like carbon) films. The shearing force can be calculated by multiplying the spring constant of the support and the difference in displacement between the cases with and without a junction (figure 5).

4.2. Shear deformation and breakage of Ag junction

Here the shear test of a silver nano junction is described [5]. Tips were covered with a 30 nm thick silver layer. The procedure of our experiment, contact-shearing-fracture process with TEM observation, is as follows:

a) Both tips were brought into contact and a junction appeared. (figures 6 a,b)
b) Shear force was applied to the junction by actuating the lateral actuator. (figure 6 c) The loading speed in the shear direction was 0.5 nm s^{-1}.
c) The junction was fractured. (figure 6 d)

When two tips met, a junction of approximately 5 nm in width appeared (figure 7 a). Although TEM images were not very clearly reproduced, we could observe lattice fringes sometime. The junction was not a single crystal. The right tip was displaced in the direction shown by the arrow, while other one was kept stationary. The junction slightly moved and tilted in the loading direction but its width stayed almost unchanged (figure 7 b). At the final stage, it became narrower down to 1-2 nm and fractured (figures 7 c, d).

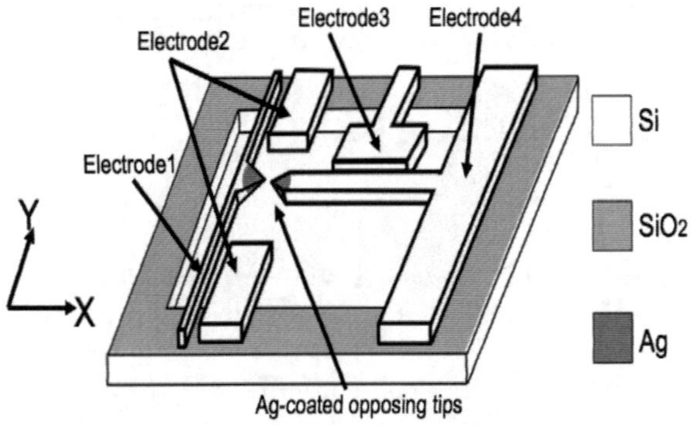

Figure 4 Schematic picture of 2-DOF MEMS device for shear testing.

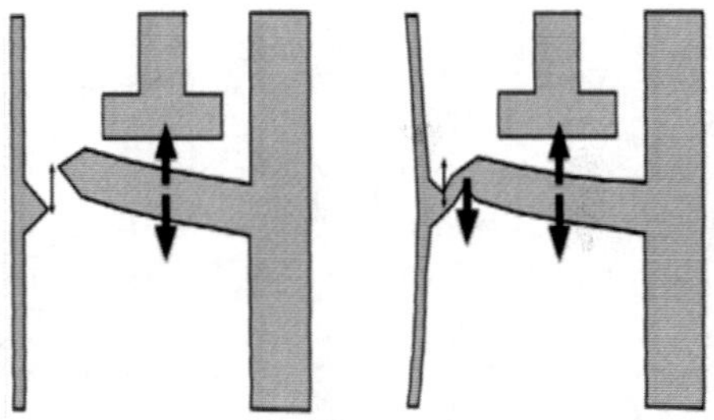

Figure 5 Schematic explanation of force measurement method.

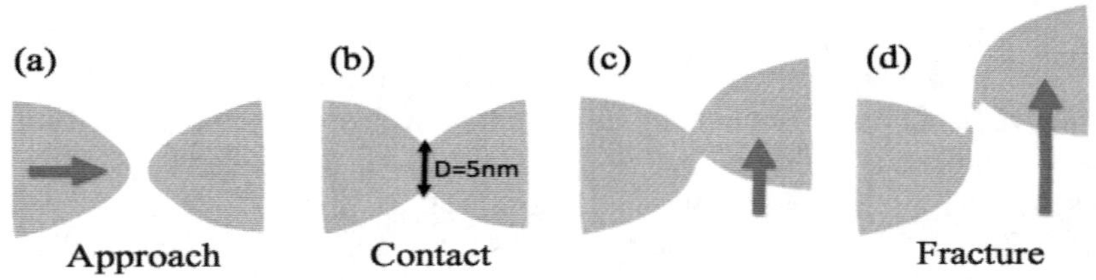

Figure 6 Schematic explanation of the experimental procedure for shear testing of nano junction.

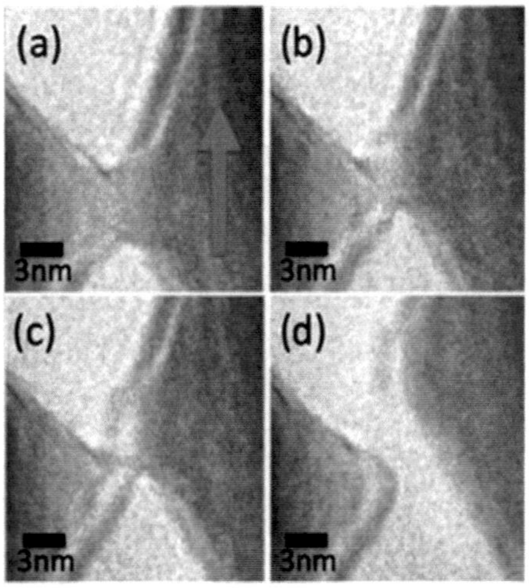

Figure 7 TEM images of silver nano junction at the different stages of deformation under shear load.

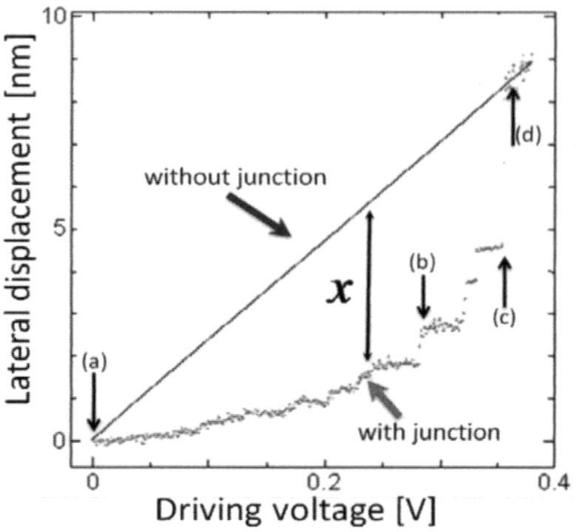

Figure 8 Tip displacement vs. applied voltage for shear testing of Ag nano junction. The arrows (a) – (d) correspond to the images of Figure 7.

Figure 8 shows the displacement vs. driving voltage characteristics with and without the junction. The one without the junction is almost linear with the sensitivity of 24 nm V^{-1}. With the junction, the movement was much smaller. At the applied voltage of approximately 0.3 V, the difference between them is 4.6 nm, corresponding to the shearing force of 10 nN. Here, the tip moved suddenly, releasing a part of the stress. The length of the stepwise motion was 1 nm. Two more stepwise motions were observed before fracture. The shearing force at the fracture was 7 nN.

4.3. Material comparison

In addition to silver tips, we also prepared tips of bare Si and those coated with Au and DLC films. The shape and dimension of nano junctions as well as the displacement of the movable tip were observed during the shear testing. The typical dimensions were 2-5 nm in diameter and 3-10 nm in length at the initial formation. Each material showed, however, very different deformation behaviors; Si junction elongated much longer than Ag and Au. The fracture forces for Ag and Au were almost the same. The fracture force for the Si junction was 10 times as large as those for Ag and Au. For the DLC interface, no junction but nano balls appeared between two tip surfaces. The ball rotated when shear load was applied to the interface. Details of those experiments were reported elsewhere [5, 6, 9].

5. Conclusion

We have developed a MEMS-in-TEM experimental system that enables the mechanical testing of nano junctions by MEMS microactuators in TEM for *in situ* atomic level observation of dynamic deformation. The applied force could also be determined. Tensile and shear tests of nano junctions were conducted for junctions made of various materials; those results showed very different characteristics among materials and from the bulk characteristics. We hope the MEMS-in-TEM system will provide new insights of mechanical properties at nano scale which are valuable to tribology study.

Acknowledgement

This work was supported by KAKENHI; the grant-in-aid of Scientific (S) 16191004 and Specially Promoted Research 21000008 sponsored by Japan Society for the Promotion of Science (JSPS).

References

[1] Fujita H (Ed.), 2003 *Micromachine as tools for nanotechnology*, (Springer)
[2] Ishida T, Nakajima Y, Kakushima K, Mita M, Toshiyoshi H and Fujita H 2010*J. Micromech. Microeng.* **20** 075011
[3] Mayer J, Giannuzzi L A, Kmino T and Michael J 2007 *MRS Bull.* **32** 400
[4] Ishida T, Kakushima K and Fujita H 2006 *IEEJ Trans. on Sensors & Micromachines* **126** 504 (in Japanese)
[5] Sato T, Ishida T and Fujita H 2010 *Proc. of Tribology Conference* p143 (in Japanese)
[6] Nabeya S, Ishida T and Fujita H 2010 *Proc. of Tribology Conference* p151 (in Japanese)
[7] Ishida T, Kakushima K, Mita M, Toshiyoshi H and Fujita H 2007 *Proc. Int. Conf. on Solid-State Sensors, Actuators and Microsystems* vol 2, p 2505
[8] Kizuka T 1998 *Phys. Rev. Lett.* **81,** 20, 4448
[9] Ishida T, Sato T and Fujita H 2010 to be presented at ICSF-2010, Mie, Japan

Nano-scale observation of frictional deformation at Ag single point contact with MEMS-in-TEM setup

Takaaki Sato, Tadashi Ishida, Shinsuke Nabeya, Hiroyuki Fujita

Institute of Industrial Science, The University of Tokyo, 4-6-1 Komaba, Meguro-ku, Tokyo, 153-8505, Japan

takaakis@iis.u-tokyo.ac.jp

Abstract. Monitoring atomic-level deformation of a frictional interface in real time is crucial for tribology research at micro and nano scale. We have combined nano-scale handling capabilities of MEMS technology with the supreme observation ability of TEM and developed a MEMS-in-TEM experimental setup. We succeeded in observing the frictional deformation of silver interface and measuring its shear force simultaneously.

1. Introduction

Scaling down the dimension of MEMS (Micro Electro Mechanical System) devices gives higher performance devices(1), sensitive sensors(2) and actuators(3). Serious problems such as friction and adhesion, however, have also been arisen because of the increase in the surface-to-volume ratio(4,5). Most of the devices such as MEMS and NEMS (Nano Electro Mechanical System) need contact points to support their body; this results in friction. The frictional phenomenon becomes a serious issue to fabricate micro miniature machines of low energy consumption and high reliability. Therefore, we need further understanding of friction and lubrication.

Conventional liquid lubricants used in macro-scale do not function at micro and nano scale because of the increased capillary force(6). Therefore, solid lubricants such as Ag(7-8), DLC and MoS_2 are good candidates of a next generation lubricant at micro and nano-scale. Since shear deformations of soft metals caused by friction easily occur at contact point, Ag is assumed to be a promising lubricant(4,6-8). However, those phenomena have not been well understood because of the experimental difficulties to observe frictional interface.

Since macroscopic frictional phenomena are attributed to the collectional behaviour of microscopic shear deformations, fundamental understanding of friction can be achieved by observing single points contact (4,5). Some groups combined TEM (Transmission Electron Microscopy) with an AFM (9-12) to observe a nano-scale deformation at the frictional interface. Those experimental systems utilize piezo actuation. The piezo actuator can produce nano motion but suffers from poor stability issues. Therefore we utilized a MEMS electrostatic actuator for a stable actuation instead of piezo actuators (13). In this research, we have observed deformation of a lubricant on a frictional interface and measured the shear force simultaneously by using MEMS-combined TEM experimental system. Observing the real area of contact under shear stress gives us a new insight in frictional characteristics and thus, we believe that the proposed experiment is crucial to reveal the lubrication mechanism.

2. Experimental setup

Our experimental setup, combination of TEM with a MEMS device was established by T. Ishida et al(14). Opposing tips of the MEMS device were actuated at room temperature to provide physical deformation at the frictional interface. The single point contact was formed between two tips and monitored by TEM to collect data for shear force measurements.

2-1 Principle of MEMS actuation

The MEMS device has 4 electrodes and two opposing tips as shown in figure.1. Electrodes provide electrostatic actuation and move the opposing tips in two directions. (1) Contact direction: By applying voltage to electrode2 in figure.1, Electrode 1 is attracted to the electrode 2. Thus, one tip is approached to the other and ended up with a physical contact. (2) Shear direction: By applying voltage to electrode 3 in figure.1, Electrode 4 is attracted to electrode 3. Thus, tip is actuated in shear direction.

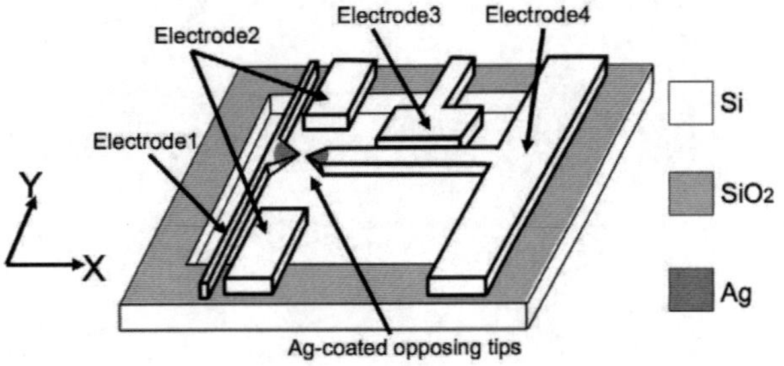

figure.1 Schematic explanation of the proposed MEMS device

2-2 Fabrication of MEMS device

A pair of suspended silicon opposing tips was made using standard lithography and double side DeepRIE process on an SOI (Silicon On Insulator) substrate. The gap between tips was defined by FIB (Focused Ion Beam) and typically 300nm. Ag was evaporated on the tips of electrodes 1 and 4(figure.1). Average deposition speed was several hundreds pm/s under 10^{-3} Pa. 20nm of silver layer was deposited only on the side-wall of the opposing tips using a shadow-mask process(14).

2-3 TEM setup

TEM (Hitachi, HF-2000UHV) was used in the experiment. The electron beam was accelerated at 200 kV and the vacuum pressure was 4.0×10^{-8}Pa. The electric current density in the observation area of the TEM electron beam was 6.2×10^{4} A· m^{-2}. With the given electron beam intensity, Joule heating was calculated to be less than 1K(15,16).

2-4 Shear force calculation

The shear force between the opposing tips can be calculated by comparing the difference of the displacements of the tips. 1) Displacement before the contact: Before bringing the opposing tips into contact, one tip was actuated in shear direction and we monitored its displacement x (figure.2(a)) which is determined by the balance between the electrostatic force, F_1, and the restoring force, F_2, of the electrode represented by equation(1). 2) Displacement after the contact: After achieving the contact, the tip was actuated in shear direction and we monitored the displacement x' (figure.2(b)) which is determined by the balance between the electrostatic force, F_1, the restoring force, F_2', and the shear force, F_3, represented by equation(2). The shear force (F_3) can be calculated as a function of x and x' by combining equations(1) and(2). Shear force can be measured by the spring stiffness and the

Journal of Physics: Conference Series **258** (2010) 012005 doi:10.1088/1742-6596/258/1/012005

difference of shear displacement monitored from the TEM images. Thus, the resolution of shear force is dependent on errors of the spring constant and the resolution of TEM images. The error of thickness, due to the DeepRIE Si etching process, is smaller than the total thickness of the electrode. Since the constant spring stiffness is proportional to the cube of the thickness, the error of the spring stiffness is negligible small compared to the effect of the resolution of TEM image. The spring constant of proposed MEMS device is 2.2 N/m and the resolution is approximately 1nm, therefore the resolution of shear force can be estimated as 2.2 nN.

$$F_1 = F_2,\ F_2 = kx \quad \ldots(1)$$

$$F_1 = F_2' + F_3,\ F_2' = kx' \quad \ldots(2)$$

$$F_3 = k(x - x') \quad \ldots(3)$$

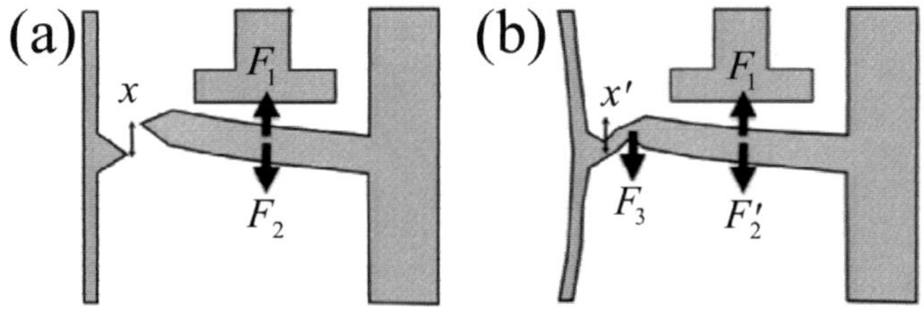

figure.2 Schematic view of the procedure to detect shear force. F_1 corresponds to the electrostatic force. F_2 corresponds to the restoring force without the junction. F_2' corresponds to the restoring force with the junction. F_3 is the reaction force to the applied shear force to the junction.

2-5. Experimental procedure

The Ag-coated silicon opposing tip (figure.3(a)) was actuated according to the experimental procedure as shown in figure.3. Tips were brought into contact forming a junction at the contact point[17] (figure.3(b)). After obtaining the contact, the one tip was actuated in shear direction with a velocity of 0.5 nm/s while maintaining the contact (figure.3(c)). The junction was finally broken by this actuation (figure.3(d)). We have monitored all steps of the experiment including the resulting shear deformation using TEM.

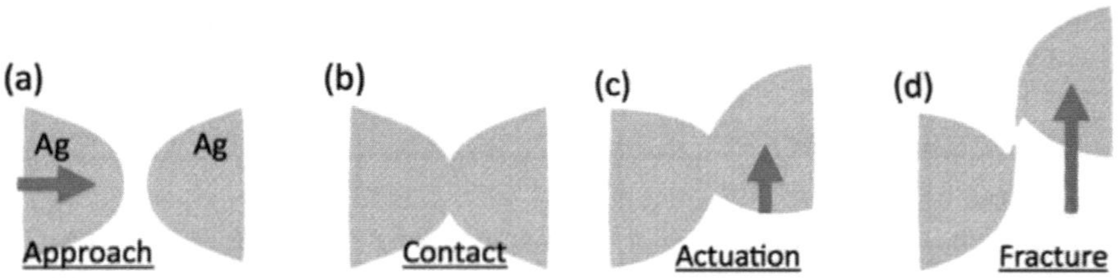

figure.3 Schematic illustration of experimental procedure for shear testing

3. Result and discussion

Changes of shear force are shown in figure.4. The lateral axis corresponds to the shear displacement and the vertical axis corresponds to the shear force. The initial junction without applied force (figure.4(a)) was actuated in the arrow direction shown in figure.5(a). The shear force increased up to 36.8nN (figure.4(d)) and the junction was finally fractured(figure.5(f)). During this shear actuation, the junction was not deformed until 10nN. The liner characteristic of the graph from 10nN to 36nN showed that the shear force changed in proportional to the shear displacement. The average gradient of the graph was about 2.9 N/m. Furthermore, some abrupt motions were observed in the region. The junction suddenly deformed for 0.6 nm in 1 second around 25nN (figure.4(b) to (c)). After the sudden deformation, the shear force built up with almost no displacement until the next sudden motion happened. Similar behaviors were observed several times before the fracture. This suggests that sudden slipping deformation is a characteristic property of Ag shear deformation.

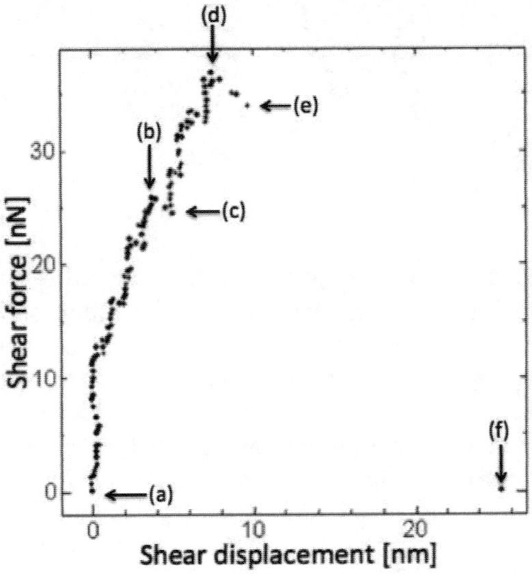

figure.4 Shear displacement vs. shear force for shear testing of Ag nano junction. The arrows (a)-(f) correspond to the images of figure.5.

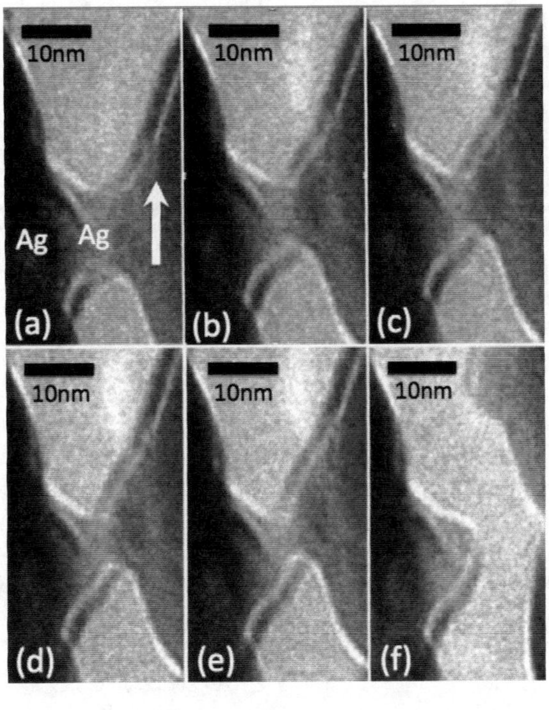

figure.5 TEM images of the shear deformation of Ag nano junction.

The size and the orientation of the junction changed during the actuation as shown in figure.5(a) and figure.5(e). To discuss the deformation quantitatively, we defined the diameter, d, of the junction and the plane angle, θ, of the cross sectional area as shown in figure.6.

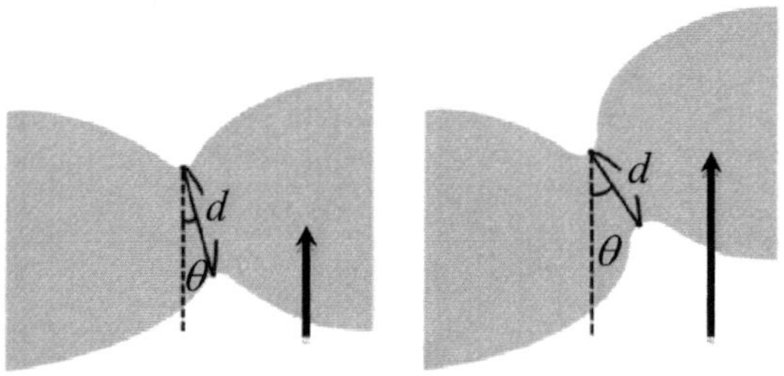

figure.6 Schematic explanation of a diameter and a plane angle, θ, of the cross sectional area

In figure.7, the lateral axis corresponds to shear displacement and the vertical axis corresponds to the angle of the cross section. The liner characteristic of the graph shows that the plane angle θ changed in proportional to the shear displacement. The initial plane angle at the beginning of the experiment (figure.7(a),5(a)) was approximately zero. The angle increases gradually with the actuation up to $27°$ (figure.7(e),5(e)) just before the fracture. Thus, the deformation by the shear actuation includes a shear deformation and a deformation in a longitudinal direction.

The change in the diameter, d, against shear displacement is shown in figure.8. The initial diameter (figure.8(a),5(a)) was 12nm. With increasing the shear displacement, the cross sectional diameter decreased linearly down to 5.8 nm (figure.8(e),5(e)) before the fracture. It is notable that there are discontinuous deformations in figure.8. The diameter discretely decreased associated with the slipping deformation in figure.4.

The sudden slipping deformation repeated several times by the shear actuation. At the moment, when the diameter of the junction discontinuously became thinner, the orientation of the junction also rotated in association.

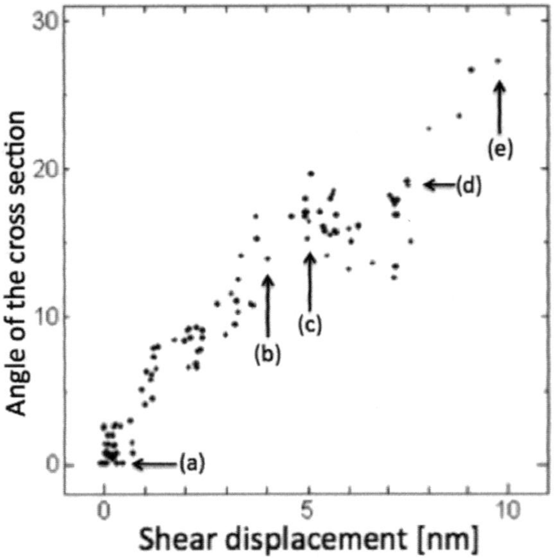

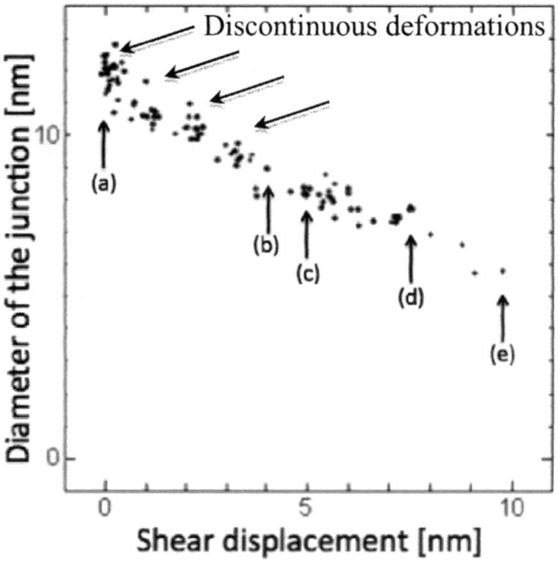

figure.7 Shear displacement vs. Plane angle of Ag junction during shear motion. The arrows (a)-(e) correspond to the images of figure.5

figure.8 Shear displacement vs. diameter curve of Ag junction during shear motion. The arrows (a)-(e) correspond to the images of figure.5

In the future we will observe that the dependency of shear force on the diameter of the junction. Furthermore, we will observe the dependency of the diameter on the normal force. Those results gives the relationship between normal force and shear force as Amonton's law(4) described. That relationship and the real time observation of shear deformation will be helpful to interpret the mechanism of the Amonton's law.

4. Conclusion

We have succeeded in observing Ag shear deformation in the nano meter range and measuring the shear force simultaneously for the purpose of revealing the mechanism of lubrication. The Ag junction discretely deformed during the shear actuation; this was associated with discrete decrease in the diameter. The plane angle of the cross sectional area rotated up to 27 degree. Thus, the deformation by the shear actuation includes a shear deformation and a deformation in a longitudinal direction. We believe that those results provide a new insight of lubrication and it is useful to reveal the mechanism of lubrication.

5. Acknowledgement

This work was supported by KAKENHI; the grant-in-aid of Specially Promoted Research 21000008 sponsored by Japan Society for the Promotion of Science (JSPS).

Reference

(1) T Hirano, T Furuhata, H Fujita , *IEICE Trans.* Electron., E 78-C (1995) **132**.
(2) C. Yamahata et al, *Biophysical Journal*, **Vol.94,** Issue 1, (2008) pp 63-70
(3) H.Fujita(Ed.), "Micromechine as tools for nanotechnology (Microtechnology and MEMS series)", Springer, 2003
(4) F.P.Bowden and D.Tabor , The friction and Lubrication of Solids, Oxford Classic Texts, (1950)
(5) B. Bhushan " INTRODUCYION TO TRIBOLOGY " John Wiley and Sons, Inc., (2002)
(6) E.-S. Yoon, R.A.Singh, H.-J. Oh, H. Kong, *Wear* **259**(2005) 1424-1431
(7) M.Goto,T.Nakahara, F.Honda Proc. IMechE, Part J:*J. Engineering Tribology* **Vol.218** (2004), pp279-291
(8) M. Goto et al, Proc. IMechE, Part J: *J. Engineering Tribology* **Vol.220** (2006), pp 135-142
(9) T.Kizuka , *Phys. Rev. Lett.* **Vol 81**,4448 (1998)
(10) Arno P. Merkle, Laurence D. Marks , *Wear* **265**(2008)1864-1869
(11) K. Anantheshwara, M.S. Bobji, *Tribology International* **43**(2010) 1099-1103
(12) Y.Kondo,K. Takayanagi, Phys. Rev. Lett, **Vol 79**(1997) pp.3455-3458
(13) T.Ishida et al, *J. Micromech. Microeng.* **20**(2010)075011
(14) T.Ishida, K.Kakushima, H.Fujita(2006) *IEEJ Trans. On Sensors and Micromachines* **126** 504 (in Japanese)
(15) S. Iijima, Microclusters, Some Experiments on Structure Instablity of Small Particles of Metals, Microcluster, edited by S.Sugano, Y. Noshona, and S. Ohnishi, Springer-Verlag, Berlin, Heidelberg(1987), pp. 186-199
(16) J. Spence, High Resolution Electron Microscopy, edited by C. Buseck and L. Eyring (Oxford University Press, New York, 1988).
(17) J.L. Coasta-Kramer, N.Garcia, P.Garcia-Mochales, P.A. Serena, Surface Science Letters, **342**(1995) L1144 - L1149

International Conference on Science of Friction 2010 (ICSF2010) IOP Publishing
Journal of Physics: Conference Series **258** (2010) 012006 doi:10.1088/1742-6596/258/1/012006

Constitutive law of dense granular matter

Takahiro Hatano

Earthquake Research Institute, University of Tokyo, 1-1-1 Yayoi, Bunkyo, Tokyo 113-0032, Japan

E-mail: hatano@eri.u-tokyo.ac.jp

Abstract. The frictional properties of dense granular matter under steady shear flow are investigated using numerical simulation. Shear flow tends to localize near the driving boundary unless the coefficient of restitution is close to zero and the driving velocity is small. The bulk friction coefficient is independent of shear rate in dense and slow flow, whereas it is an increasing function of shear rate in rapid flow. The coefficient of restitution affects the friction coefficient only in such rapid flow. Contrastingly, in dense and slow regime, the friction coefficient is independent of the coefficient of restitution and mainly determined by the elementary friction coefficient and the rotation of grains. It is found that the mismatch between the vorticity of flow and the angular frequency of grains plays a key role to the frictional properties of sheared granular matter.

1. Introduction

Granular flow is ubiquitous in solid earth sciences and engineering: e.g. avalanches, landslides, debris flow, silo flow, etc. These phenomena are essentially dominated by the frictional properties of granular matter. Thus, to find a law that describes the behavior of friction coefficient (ratio of the shear stress to the normal stress) of granular matter is an essential problem [1].

The mechanical properties of granular matter generally depend on many ingredients: density, shear rate, pressure, temperature, humidity, interstitial fluid, etc. These ingredients must be carefully controlled in investigating the frictional properties, as they affect the result in unexpected ways. Because such potentially important ingredients are not known a priori, despite extensive experimental efforts, the frictional properties of dense granular matter are still not clear.

In order to rule out chemical processes that potentially affect the frictional properties, numerical simulation plays an important role. Through extensive simulations [2,3], it turns out that the behavior of friction coefficient of dry granular matter is well described by a nondimensional number, $I = \dot{\gamma}\sqrt{m/Pd}$, where $\dot{\gamma}$ denotes shear rate, m is the mass of a grain, P is the normal pressure, and d is the grain diameter. This nondimensional number is referred to as the *inertial number*, which seems to be successful in stating a constitutive law for fast granular flow [4].

To clarify a constitutive law for slow granular flow, da Cruz et al. performed an extensive simulation on a two dimensional system in a wide range of the inertial number: $6 \times 10^{-4} \leq I \leq 0.3$. They find a friction law that reads $\mu = \mu_0 + aI$, where μ is the kinetic friction coefficient, and μ_0 and a are positive constants. However, in three dimensional systems, somewhat different friction law is found, where the inertial-number dependence of friction coefficient is described

Journal of Physics: Conference Series **258** (2010) 012006 doi:10.1088/1742-6596/258/1/012006

by a power law, which is of the same form as the Herschel-Bulkley model; i.e., $\mu = \mu_0 + sI^{\phi}$, where s is a positive constant and $\phi \simeq 0.3$ [5,6]. This power law behavior is confirmed in a numerical simulation of a certain class of granular matter: particularly smooth (frictionless) beads. However, the validity of this power-law friction in a wide class of granular materials, as well as the theoretical understanding and derivation, are still open. In addition, before theoretical consideration, some questions regarding phenomenology are still open: How do the coefficient of restitution, rolling resistance, and other material properties affect the frictional properties of granular matter? In this paper, these questions shall be answered.

2. Model

2.1. *force model*

Here each grain is assumed to be sphere. The interaction between grains is described by the discrete element method (DEM) [7], which is the standard model used in powder engineering and soil mechanics. Consider a grain i of radius R_i located at $\mathbf{r}_i$ with the translational velocity $\mathbf{v}_i$ and the angular velocity $\mathbf{\Omega}_i$. This grain interacts with another grain j whenever they are in contact; i.e., $|\mathbf{r}_{ij}| < R_i + R_j$, where $\mathbf{r}_{ij} = \mathbf{r}_i - \mathbf{r}_j$. The force acting on two particles is decomposed into two direction: normal and transverse to $\mathbf{r}_{ij}$. Introducing the normal unit vector $\mathbf{n}_{ij} = \mathbf{r}_{ij}/|\mathbf{r}_{ij}|$, the normal force $\mathbf{F}_{ij}$ is given by $[kh_{ij} - \zeta \mathbf{n}_{ij} \cdot \dot{\mathbf{r}}_{ij}]\mathbf{n}_{ij}$, where $h_{ij} = R_i + R_j - |\mathbf{r}_{ij}|$ and k is a constant. The coefficient of restitution, e, can be calculated using the relation

$$e = \exp\left[-\frac{\pi}{\sqrt{4\bar{m}k/\zeta - 1}}\right], \tag{1}$$

where $\bar{m} = m_i m_j/(m_i + m_j)$. In order to define the transverse force, we define the relative tangential velocity as

$$\Xi_{ij} = \dot{\mathbf{r}}_{ij} - \mathbf{n}_{ij} \cdot \dot{\mathbf{r}}_{ij} + \frac{R_i\mathbf{\Omega}_i + R_j\mathbf{\Omega}_j}{R_i + R_j} \times \mathbf{r}_{ij}$$

and introduce the relative tangential displacement vector Θ as $\int_{roll} ds\Xi(s)$. The subscript of the integral indicates that the integral is performed only when the contact is *rolling*; i.e., $k_t|\Theta_{ij}| < \mu_e|\mathbf{F}_{ij}|$ or $\Xi_{ij} \cdot \Theta_{ij} < 0$. Otherwise, the contact is said to be *sliding*. The magnitude of the tangential force depends on the state of the contact: $\mu_e|\mathbf{F}_{ij}|$ for sliding contact and $k_t|\Theta_{ij}|$ for rolling contact.

2.2. *configuration*

The particles are bidisperse: the diameters are $0.8d$ and $1.0d$, respectively. The number of each particle is the same. For simplicity, the mass of the grains is set to be the same, which is denoted by m.

The dimensions of the system are $L \times L \times H$, where we use periodic boundary conditions along the x and the y axes. In the z direction, there exist two rigid walls that consist of the larger particles. One of the walls is displaced along the x axis at constant velocity V to realize plain shear flow, where the velocity gradient is formed in the z direction. The grains that constitute walls interact with the bulk grains via the force described above. This wall is also allowed to move along the z axis so that the pressure is kept constant at P, while it is immobile along the y axis. Namely, the z dimension of the system, denoted by H, depends on the velocity of the wall. Here we define that H_0 is the layer thickness at $V = 0$. Here the dimensions are set to be $H_0 = 12d$ and $L = 10d$, in which approximately $2,000$ particles are intervened between the walls. See Fig. 1 for schematic of the computational system. The equation of motion of the wall along the z axis is given as $M\ddot{H} = F_z - PL^2$, where M denotes the mass of the wall and F_z is the repulsive force given by the grains. The mass of the wall, M, is set to be $M = 100m$. It should be remarked that the gravity is not taken into account, because the pressure P shall be chosen to be large enough.

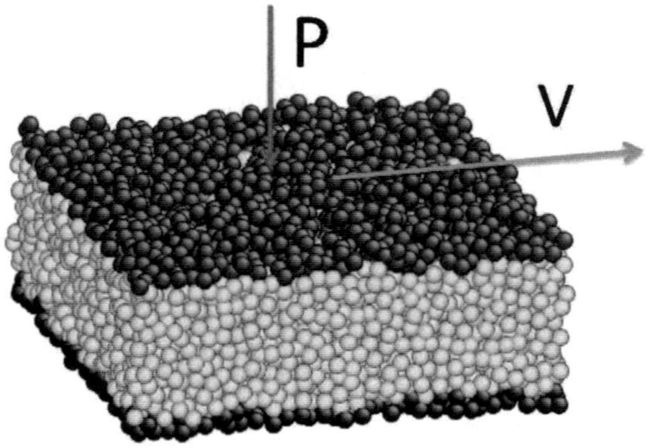

Figure 1. Schematic of the computational model.

2.3. parameters
The transverse spring coefficient is set to be $k_t = k/5$ throughout this study. This relation is empirically known to reproduce experimental results. The other parameters, μ_e and ζ, generally depends on the material of grains. It is thus important to investigate how these parameters affect the frictional properties of granular matter. The pressure P is kept fixed as $Pd/k = 8.6 \times 10^{-4}$. For de-dimensionalization, we set $d = 1$, $k = 1$, and $m = 1$. Here we estimate the material constants based on the sound velocity and the Young's modulus, which roughly corresponds to $d\sqrt{k/m}$ and k/d, respectively. (Here the numerical factors are neglected.) Note also that any length scale in the simulation is scaled with the grain diameter, d. Thus, the unit velocity and the unit pressure in DEM are of the order of kilometer per second and several tens of Gigapascal, respectively. The pressure applied to the system is thus on the order of 10 MPa. In this paper, we limit ourselves to the frictional property of steady states. The steady state is realized after certain strain, which can be confirmed by observing the friction coefficient and the height of the wall, H. Here the computational data is taken only after the system is subject to 100 % strain. The traction acting on the moving wall is monitored, which is denoted by F_x, so that the bulk friction coefficient of the system is defined as F_x/PL^2.

3. Velocity profile

In a certain class of systems including granular matter, the uniform shear deformation can be unstable and deformation tends to localize within a narrow band. This spontaneous structure formation is referred to as shear-banding. Although it is speculated that some healing processes (or thixotropy, velocity-weakening friction, etc) may play a certain role in shear-banding [8], the general mechanism for shear banding is still not clear.

Thus, the internal flow structure (velocity profile) should be measured before discussing the frictional properties. In order to exclude the fluctuation, the instantaneous flow velocity is averaged over a certain strain (100 %). Fig. 2 shows the averaged flow velocity as a function of the depth. Although any healing mechanism is not modeled in the present computational system, the behavior that is similar to shear-banding is observed. Indeed, shear flow tends to localize near one of the wall. It is found that the flow localization is more apparent for systems with larger coefficient of restitution (i.e., with smaller damping coefficient ζ.)

The flow localization is also affected by the driving velocity of the wall. In the right panel of Fig. 2, where the effect of damping is significant ($\zeta = 1$), the shear flow is uniform except for the highest velocity case, $V = 10^{-1}$. Thus, flow localization is more likely to occur at larger

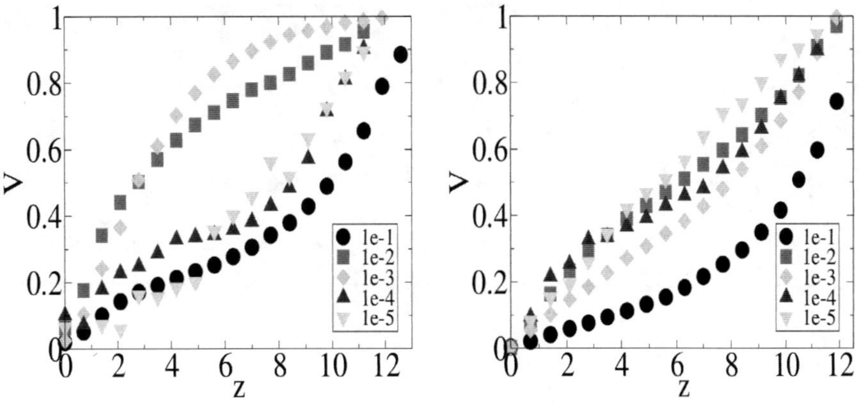

Figure 2. Flow velocity profiles along the depth at various wall velocities. The legends denote the wall velocity. The flow velocity is normalized by the maximum velocity (the wall velocity). (Left) $\zeta = 0.01$, (Right) $\zeta = 1.0$.

driving velocity of the wall.

It should be remarked that the localization near the wall is also observed in a two dimensional system [9], where the localization is more frequently observed in a system at lower pressure. This tendency is also observed in the present three-dimensional system.

4. Behavior of the bulk friction coefficient

4.1. general description

Shear stress σ in steady shear flow is determined by the balance between power input and the dissipation rate D.

$$\dot{\gamma}\sigma = D \tag{2}$$

where $\dot{\gamma}$ denotes shear rate. Thus, the nature of shear stress (and thus friction coefficient) involves the mechanism of dissipation. Note that the present computational model has two different sources of dissipation at the particle level: damping in the normal force and friction in the tangential force. The former is proportional to the damping coefficient ζ and the latter is proportional to the elementary friction coefficient μ_e. Therefore, the dependence of friction coefficient on each parameter may be a key to the understanding the dissipation mechanism in sheared granular matter.

4.2. shear rate dependence

First, the shear rate dependence of the friction coefficient is discussed focusing the effect of the damping coefficient ζ. Fig. 3 shows the bulk friction coefficient as a function of the inertial number I (i.e., nondimensional shear rate) with several values of ζ. Here three cases are investigated: $\zeta = 10^{-2}, 10^{-1}, 1$, whereas the elementary friction coefficient is fixed at 0.6. In terms of the coefficient of restitution e, these values of correspond to $e = 0.80$, 0.49, and 0.043, respectively. (Note Eq. (1)). The bulk friction coefficient is insensitive to shear rate and to the damping coefficient in the lower I regime $I \leq 10^{-2}$. This indicates that the collision in the normal direction is irrelevant to the bulk friction coefficient in dense and slow granular flow. Note also that it is much less than the elementary friction coefficient.

The bulk friction coefficient then sharply increases around $I \sim 10^{-1}$. In such high I region, the friction coefficient apparently depends on the coefficient of restitution: it increases as the coefficient of restitution decreases. (For example, at $I = 0.28$, $\mu = 0.44$ for $e = 0.80$, $\mu = 0.50$

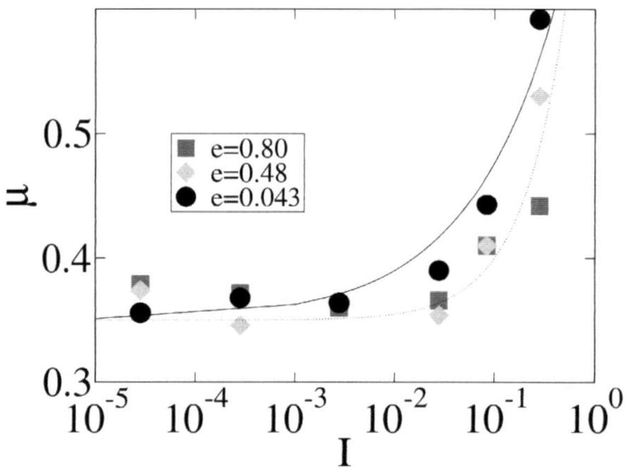

Figure 3. The friction coefficient as a function of the inertial number defined by Eq. (1). The legends denote the coefficient of restitution. The solid and dotted lines are empirical constitutive laws, Eq. (3), where $\phi = 0.5$ for the solid line and $\phi = 1.0$ for the dotted line.

for $e = 0.49$, and $\mu = 0.57$ for $e = 0.043$.) This implies that the dissipation due to normal collision is dominant in the rapid flow region where $I \geq 10^{-1}$. The shear rate dependence can be empirically described by the Hershel-Bulkley model,

$$\mu = \mu_0 + sI^{\phi} \tag{3}$$

Here the exponent ϕ ranges from 0.5 to 1.0. (See Fig. 3). It should be remarked that these values are larger than previously estimated [5,6], where $0.3 \leq \phi \leq 0.4$. The difference may be due to the elementary friction coefficient. In the previous studies [5,6], the elementary friction coefficient is set to be zero or 0.2, while it is 0.6 here. Thus, the exponent ϕ depends on the elementary friction coefficient. Although $\mu_e = 0.6$ case is also investigated in Ref. [5], the data are insufficient for the estimate of the exponent.

4.3. effect of elementary friction coefficient

In dense and slow granular flow, the elementary friction coefficient between grains, μ_e, is more important parameter than the coefficient of restitution. Fig. 4 shows how the bulk friction coefficient of a slowly-sheared system ($I = 2.8 \times 10^{-5}$) depends on elementary friction coefficient. The bulk friction coefficient increases as the elementary friction coefficient increases at smaller μ_e, but then becomes independent of μ_e. (There is a maximum value around 0.35). It should be noted that this insensitivity of the bulk friction coefficient to the elementary friction coefficient is also found in two dimensional granular systems [3].

5. Rotation of grains

As shown in Figs. 3 and 4, in dense and slow flow, the extent of dissipation in normal interaction (i.e., the coefficient of restitution) is irrelevant to the bulk friction coefficient but sliding friction between grains plays an important role. Now let us estimate the bulk friction coefficient in the low I regime taking the sliding friction between grains into account.

For simplicity, we consider the two dimensional case. Neglecting the fluctuation in the translational velocity of each particle, the sliding velocity at the contact of two particles is approximately

$$d\,(\dot{\gamma}\cos\theta_{ij} - (\Omega_i + \Omega_j)/2) \tag{4}$$

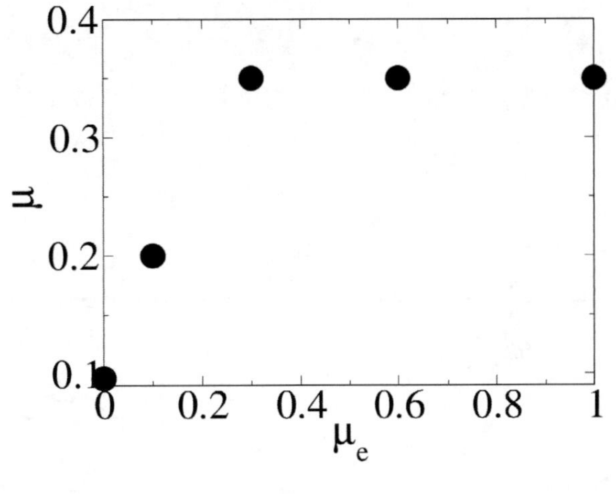

Figure 4. The bulk friction coefficient taken at $I = 2.8 \times 10^{-5}$ (the slowest data in Fig. 3) as a function of the elementary friction coefficient.

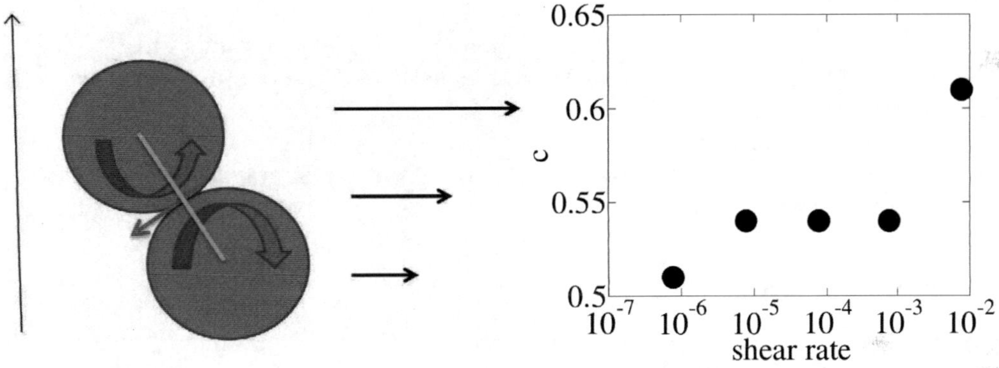

Figure 5. (Left) Schematic of grain rotation and shear flow. (Right) The frustration parameter as a function of shear rate. Here $\zeta = 1.0$ and $\mu_e = 0.6$.

where $\cos\theta_{ij} = \mathbf{n}_{ij} \cdot \mathbf{n}_z$. (See the left panel of Fig. 5). Note that $\theta_{ij} \leq 0$ without the loss of generality. For simplicity, the diameter is assumed to be the same for the all particles. Using the average normal force, $Pn_b d$, where n_b is the number of contacts in unit volume, the frictional force acting between particles i and j is $\mu_e P/n_b d$. Then the energy dissipation rate per contact reads

$$\mu_e P n_b^{-1} |\dot{\gamma} \cos\theta_{ij} - (\Omega_i + \Omega_j)/2| \quad (5)$$

Using the power balance Eq. (2), the bulk friction coefficient is estimated as

$$\mu = \mu_e \left\langle \left| \cos\theta_{ij} - \frac{\Omega_i + \Omega_j}{2\dot{\gamma}} \right| \right\rangle, \quad (6)$$

where the bracket denotes the average over the contacts. Equation (6) indicates that the friction coefficient vanishes if the grains have their neighbors only in the velocity gradient direction (thus $\cos\theta_{ij} = 1$) and $\Omega_i = \dot{\gamma}$. Similarly, the friction coefficient can also vanish if the grains have their neighbors only in the flow direction (thus $\cos\theta_{ij} = 0$) and $\Omega_i = \Omega_j$. However, the orientations

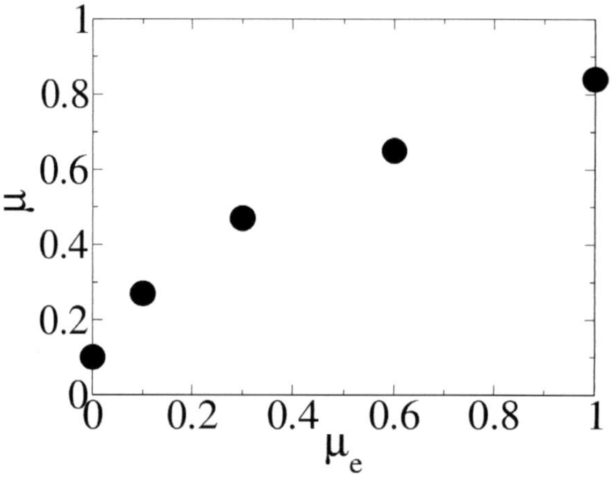

Figure 6. The bulk friction coefficient taken at as a function of the elementary friction coefficient between particles. Here the grains are not allowed to rotate.

of contacts in granular matter are of course random. This is a reminiscence of spin-glasses; the interaction in the flow direction is antiferromagnetic and the one in the gradient direction is ferromagnetic. In addition, there are intermediate interactions and these interactions are coupled to translational velocity.

As the problem is very complicated, let us make further simplification that the angular frequency of particles are the same, i.e., $\Omega_i = \omega$ and $\cos\theta_{ij}$ is replaced by unity. Then Eq. (6) becomes

$$\mu \sim \mu_e \left\langle \left| 1 - \frac{\omega}{\dot{\gamma}} \right| \right\rangle \tag{7}$$

The quantity $1 - \omega/\dot{\gamma}$, which we shall refer to as the frustration parameter, is thus important to the nature of friction in dense and slow granular flow. The right panel of Fig. 5 shows the typical behavior of the frustration parameter c as a function of shear rate. In slow and dense flow regime, the frustration parameter takes the value around 0.5. This means $\mu \sim 0.5\mu_e$, which is not a very bad prediction for $\mu_e = 0.6$ case, where $\mu \simeq 0.35$. (See Fig. 4). Note that the value of frustration parameter may depend on μ_e. At this point, we do not know how this quantity is determined from the contact network of grains.

In order to test the validity of Eq. (7), one can consider a model in which the grains do not rotate at all. This leads to $\omega = 0$ so that it is expected from Eq. (7) that$\mu \simeq \mu_e$. Fig. 6 shows the bulk friction coefficient of this model, which increases as the elementary friction coefficient increases up to 1.0. This behavior makes a quite contrast to that of the previous model with the particle rotation.

6. Conclusions

The friction coefficient of dense granular matter is an increasing function of shear rate. The coefficient of restitution affects the friction coefficient only in rapid flow. Contrastingly, in dense and slow regime, the friction coefficient is mainly determined by the elementary friction coefficient and the rotation of grains. Mismatch between the vorticity of flow and the angular frequency of grains plays a key role to the friction coefficient.

References

[1] C. Marone, Annu. Rev. Earth Planet Sci. **26**, 643 (1998)

[2] GDR MiDi, Eur. Phys. J. E, **14**, 341 (2004)
[3] F. da Cruz, S. Emam, M. Prochnow, J.-N. Roux, F. Chevoir, Phys. Rev. E, **72** 021309 (2005)
[4] P. Jop, Y. Forterre, O. Pouliquen, Nature **441** 727 (2006)
[5] T. Hatano, Phys. Rev. E **75**, 060301 (2007)
[6] P.-E. Peyneau, Roux J.-N., Phys. Rev. E, **78** 011307 (2008)
[7] P. A. Cundall, O. D. L. Strack, Geotechnique, **29** 47 (1979)
[8] L. Isa, R. Besseling, and W. C. K. Poon, Phys. Rev. Lett. **98**, 198305 (2007)
[9] E. Aharonov and D. Sparks, Phys. Rev. E, **65**, 051302 (2002)

Journal of Physics: Conference Series **258** (2010) 012007 doi:10.1088/1742-6596/258/1/012007

Friction of the short model ski at low velocity

Maiko Takeda[1], Kunio Nikki[1], Takaaki Nishizuka[1] and Osamu Abe[2]

[1] Department of Applied Physics and Chemistry, the University of Electro-Communications, 1-5-1 Chofugaoka, Chofu-shi, Tokyo 182-8585, Japan
[2] Snow and Ice Research Center, National Research Institute for Earth Science and Disaster Privention, Tokamachi, Sinjo, Yamagata 996-0091, Japan

E-mail: takeda@nikki.pc.uec.ac.jp

Abstract. Frictional coefficients of a model ski (10 cm length) were measured at low velocity (0.005 ~ 1 m/s) by means of the tribometer method. The coefficient values depend on sliding velocity, snow temperature and snow grain size. No load effect was observed for any temperature. The static frictional coefficient was depended on the logarithm of the contact time. Further, the coefficient increased with the temperature going up for the smaller grain snow. These experimental results show that the origin of the friction force of the model ski at low velocity is the shearing force of the adhesion between snow and polyethylene.

1. Introduction

It is well known that the kinetic frictional coefficient (μ) of ski and ice is exceedingly small in comparison with μ of the usual solid-solid contact. There are two hypothesises why ski slides well. One is the adhesion theory that the shearing strength of ice is small in comparison with Brinell hardness [1, 2]. As an example, the friction on (1-10) plain of ice crystal was about two times greater than on (0001) [3]. Another is the melt-water lubrication theory. The water which works as a lubricant is generated with the frictional heat between the snow and the ski [2, 4-8]. For example, the melt-water has been observed in a laboratory experiment for an interface between the slider sole and the ice where the frictional heat is accumulated [7]. The temperature rises of the sole were observed near the melting point (-10 to 0 °C) by thermocouple when a down-hill ski glided with 2 to 10 m/s. The upper most temperature of the gliding ski base was dependent on the ambient temperature [9]. Recently, the influence of the snow temperature (the lowest of μ appeared at around -3 °C) under racing conditions was emphasized [10]. Additionally, the gliding theory other than the melt-water lubrication was proposed to understand the ski gliding in various conditions. A lot of works related to the snow friction were reviewed and discussed by Colbeck [11].

When the ski starts to glide, the temperature of the snow surface which contacts with the ski is heated with the friction and the heat is accumulated. Viewed from the ski, the heat is accumulated for the ski sole by friction with the snow surface, but the top part of the sole is always cooling by the coming new snow and air. Then the temperature of ski sole would be dependent on the distance from the ski top. The frictional condition is also different with the sole position. The heat accumulation also depend on the heat conduction of ski materials, the snow condition, ambient temperatures and so on. Only an average μ value of a sliding ski could be obtained by experiment. It is impossible to analyze the friction state at an individual sole position by changing the friction parameters on experiment. It is also difficult to estimate the real contact area of the gliding skis. We can observe temperatures of a ski sole but cannot observe temperatures of the surface of snow particles, rubbed by the ski. As the μ value of a short slider is equivalent to that of the head part of the actual ski, the mechanism of the friction would be simple and the heat accumulation would not occur especially for the slow velocity.

In this paper, nearly the intrinsic frictional resistance of a short model ski at low velocity was observed by means of the tribometer method with carefully prepared hard packed flat snow in the Cryospheric Environment Simulator of NIED with artificial snow and in the fresh air of the Sugadaira Observatory of the University of Electro-communications (UEC) in Nagano, with a variety of natural snow. The snow temperature, the gliding velocity and the load effects of μ were observed and the mechanism of the frictional resistance was discussed. In this paper, the low velocity expresses 0.001 to 1 m/s, the low temperature means -10 to -15 °C, and the high temperature is -5 to -1 °C.

2. Experiments

2.1. The tribometer method

A schematic diagram of the tribometer is presented in Fig. 1. As the diameter of the turn-table is 180 cm, the snow track can be regarded as a pseudo-straight track for the short model ski (length: 10 cm, width: 5.5 cm) at low gliding velocity (the centrifugal force of friction is negligibly small). The slider "S" is strapped to a static frame "L" via a load cell and the large snow-table is rotating under the model ski by an electric motor with a speed controller. Snow surface of the track on the table is prepared by a flattening blade "FB" which is fixed to the static frame "L" as horizontal accurately. The apparent μ is easily deduced by measuring the traction force "T" of the slider, by using the load cell. ($T = \mu \times$ mg) The actual track rotation angle and the velocity are measured with a rotary encoder "R" located at the center of the table.

2.2. The Model Ski

The model ski was a remake of an actual alpine ski with polyethylene-sole. It was cut for a certain length and the sliding surface made flat by means of the insertion of a duralumin plate (10 mm in thickness) on the sole polyethylene. The length and the width of a model ski are 10 and 5.5cm, respectively. The weight is changeable to be four times. No wax was used.

2.3. The Snow Preparation

Artificial snow which made in the Cryospheric Environment Simulator of NIED and a variety of natural snow at Sugadaira in Nagano were used as sample snow. The grain size of snow was selected by sieves. The prepared snow was placed on the turn table of the tribometer and the snow surface made flat.

Snow hardness is an important factor to discriminate the energy loss of a ski sinking into the snow from the sliding resistance, but it is not easy to control the surface one [12]. On the other hand, snow

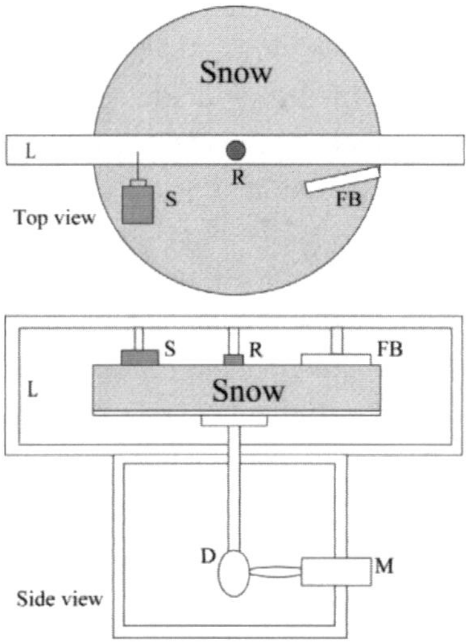

Figure 1. A schematic diagram of the tribometer (L: Static frame, FB: Flattening blade, S: Slider, R: Rotary encoder, D: Gear box, M: Motor).

density is easy to control and seems an important factor correlating to the effect of ski sinking into the snow. In this paper, the snow densities of the test snow track were prepared to be the same for a series of experiments. The density of the sample snow was fixed to be 360~550 kg/m^3. Before the experiment, the snow was sintered for 6 to 16 hours at the observed temperature to harden the surface.

In addition, the surface hardness of the test snow was observed by a homemade penetrating type of hardness meter. Indented depth was about 2 mm (diameter of the pressure head was 1 cm). The pressure was around 2 kg/cm^2. The pressure of the contact area of the model ski was about 4.8g / cm^2 (1/2 of the actual ski). It may be given as a conclusion that the surface snow condition was hard enough to compare to the actual ski gliding, which indicates that the pack and plow resistances are negligible.

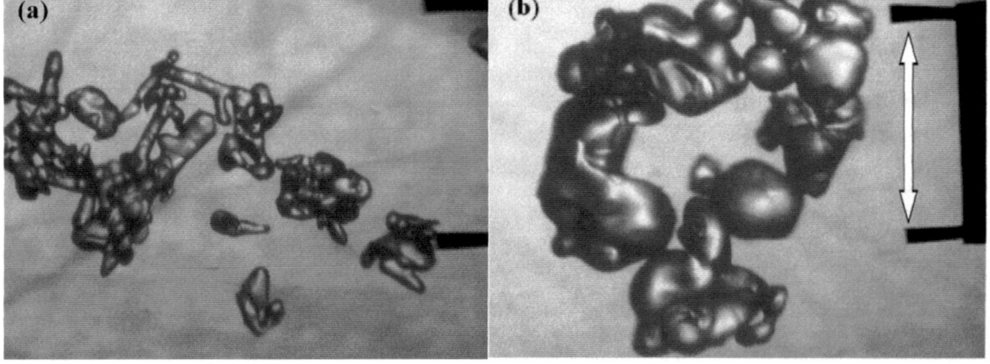

Figure 2. The snow grains of packed snow surface after sintering. The arrow corresponds to 1 mm in length for both pictures.

3. Experimental Results and Discussions

3.1. Velocity dependence of the kinetic frictional coefficient (μ)

The kinetic frictional coefficient (μ) *vs.* the sliding velocity for artificial snow at several temperatures is shown in Fig. 3. It suggests that there are several mechanisms of gliding friction working simultaneously or separately. The μ value was decreasing with the decreasing snow temperature until -10 °C at low velocity (lower than 3cm/s). The μ values were exceptionally smaller than the friction between usual solids. The μ value at around -10 °C (ca. 0.05) was smallest and it is close with the value of the best condition of the actual alpine ski at -3 to -5 °C [10]. The μ value of fast slides (higher than 10 cm/s) is decreased with the increase of temperature. Namely, the sliding at low velocity showed the high μ value but the high velocity sliding showed the low μ value at only the high temperature. These results clearly show that the temperature dependency of μ is drastically changed with the sliding velocity at high temperature. As a sliding at higher temperature and the faster, the velocity seems to easily generate the melt-water, and it looks reasonable that the feature of μ can be explained by means of the "well known" melt-water lubrication hypothesis.

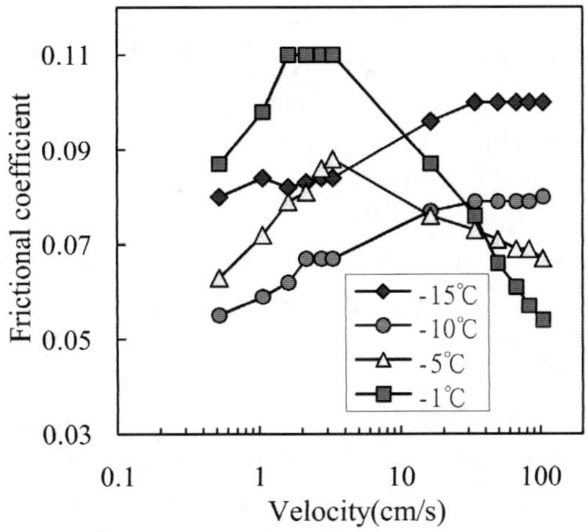

Figure 3. μ vs. the sliding velocity for an artificial snow at several temperatures. The snow grain size is 0.1mm in Fig. 2 (a).

3.2. Normal force effect of the kinetic frictional coefficient (μ)

According to the melt-water lubrication theory, it is reasonable that μ is expected to be small when the velocity and/or normal forces are increased. It is frequently quoted in the ski science field that the main mechanism of the friction of ski at higher the velocity than 10 m/s is explained by the lubrication of the melt-water which is generated by the frictional heating [4].

The frictional heat (Q) expressed as follows;

$$Q = \mu M v. \quad (1)$$

Here, M is the normal force, v is the gliding velocity and μ is the frictional coefficient. It is clear from (1), when the normal force and/or the velocity are increased, the frictional heating and then the thickness of water film would be increased and then the friction force would be decreased. In fact, the lowest of μ appeared at around -3 °C for actual ski gliding at racing condition [10]. Velocity dependences of μ accompanied with the load change of the model ski at the different temperatures are shown in Fig. 4. The load was altered by adding metal plate attachments.

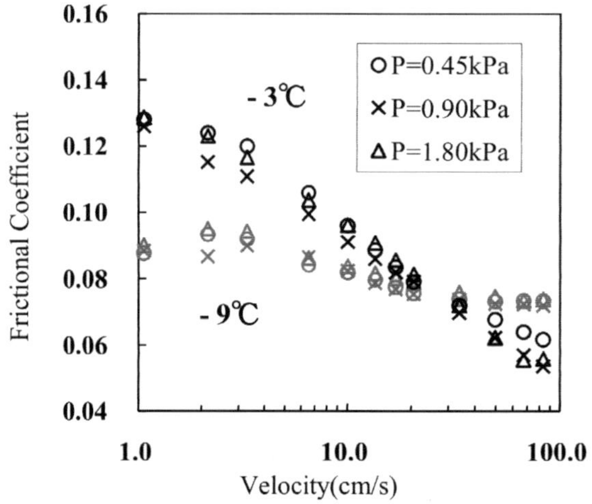

Figure 4. Velocity dependence of μ using different load (○: usual weight, ×: twice, △: forth) of the ski at several temperatures with natural snow. (The tribometer method in fresh air of the Sugadaira Observatory)

Figure 4 shows that μ decreased when the gliding velocity was increased, but an obvious normal force effect was not observed even when the normal force became twice or four times any temperature. Further, the μ at lower velocity was larger at higher temperature as the same with Fig. 3. It is unlikely that the melt-water works as lubricant at around 1 m/s for any temperature.

3.3. The contact time dependence of the static frictional coefficient (μ_s)

In Fig. 5, the static frictional coefficient (μ_s) *vs*. the contact time of the different snow grain size is shown at several temperatures. First of all, μ_s depended on the logarithm of the contact time for both snows. The μ_s value increased with contact time. The reason is that the real contact area increases with going time [13, 14]. The contact time dependence of μ_s affects the velocity dependence of μ. The real contact area decreases with increasing velocity because the contact time is short at high velocity [15]. The friction mechanism would be explained by the adhesion theory. Second, the real contact area would be larger for smaller grain size, because the number of contact positions seems bigger for smaller grain size. In the figure, the μ_s values were larger for the small grain. It reveals that the real contact area is larger for the small grain. Further, the contact time dependence of μ_s in Fig. 5 (a) at the high temperature is larger than the low temperature and μ_s in Fig. 5 (b). The friction force increased very much at high temperature for many contact points case (grain size 0.1mm), and did not increase so much for smaller contact points (grain size 0.5).

These results suggest that the contact between ski and snow is a dry process. Further, it is clear that the temperature effect was far larger for the small grain. This is a new finding. For the explanation, we introduce the existence of the liquid-like layer on the surface of snow grain. The surface energy of liquid is larger than that of solids. So the adhesion force is larger for liquid-like layers than for solid ice. As a result, it is reasonable that the adhesion force would be enforced by the liquid-like layer. Also, the thickness of the liquid-like layer of snow crystal increased with increasing temperature at premelting point. Then the adhesion force was increased near the melting temperature of the ice.

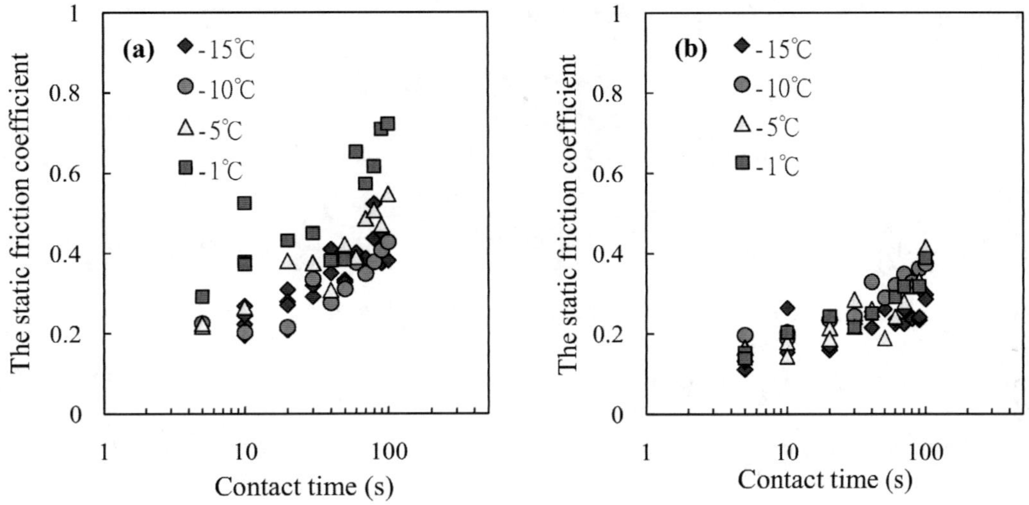

Figure 5. μ_s vs. the contact time. Each snow grain sizes are (a) 0.1mm and (b) 0.5mm.

3.4. Snow grain size effect of the kinetic frictional coefficient (μ)

We measured the coefficient of bigger grain size (0.5 mm) in Fig. 6. In consequence of the experiment, the μ value of the bigger grain is smaller than that of the smaller one (Fig. 3). The interesting velocity dependence of the frictional coefficient for the high temperature in Fig. 3 disappeared. But at premelting point, the μ value of high velocity showed the same value with the small grain. This is reasonable because the friction of high velocity does not have the effect of adhesion so much because the contact area is small. These results are concerned with the smaller contact points, and then the smaller adhesion force.

3.5. Discussion of the origin of the friction force (μ)

It is known that the frictional resistance between two ices is generated by the adhesion at the contact surfaces when the gliding velocity is slower than 10 cm/s. The friction force is regarded as the plastic shear deformation force of the thin film of ice between the contact surfaces [3, 16, 17]. To connect the frictional resistance of ice, it was reported that the μ value of a steel ball on a single crystal of ice at very low gliding velocity (which means no melt-water generation) became a very small value (0.005 to 0.2). The μ value on the prismatic plain was twice as large as that on the basal plane and there was no normal force effect, which is explained by the adhesion theory [3]. According to the well-known melt-water lubrication hypotheses, it is unlikely that the melting points are different with the crystal axis of the same crystal.

In the ski sliding, interaction between polyethylene and snow, it seems that the attractive interaction

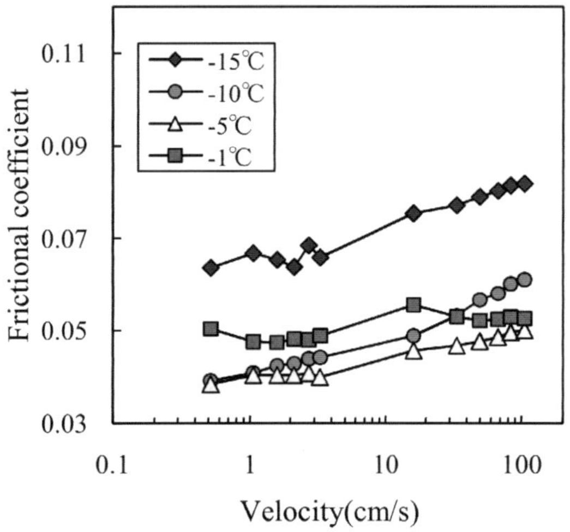

Figure 6. μ vs. the sliding velocity for a snow at several temperatures. The snow grain size is 0.5 mm in Fig. 2 (b).

force would arise from the dipole-induced dipole interaction. The dielectric constant of liquid water is far larger compared to the solid ice, because of the mobility of water molecules is far larger for liquid water. The attractive interaction energy of polyethylene is expected to be larger with liquid-water than with solid-ice.

Also, it is well-known that the thickness of the liquid-like layer of snow crystal increased with the increased temperature close to the melting point [18]. It accounts for the adhesion effect increased at high temperature and μ_s in Fig. 5 (a) was more than the low temperature. In this case, the gliding resistance is consistent with the shearing force of the adhesion between the snow and the polyethylene sole and means a plastic deformation of the liquid-like layer at the contact surface.

As our experimental results indicate, μ showed a very small value at around -10 °C for very slow slides (Fig. 3). In this temperature and sliding velocity range, it is likely that no melt-water is expected. This result would be connected with the very small adhesion force between the polyethylene slider and snow. It is reasonable that as the liquid-like layer becomes thin, then the adhesion force also becomes small.

On the other hand, the μ value was increased with decreasing the temperature from -10 °C. For this condition, the liquid-like layer practically disappeared [18], and at the same time the hardness of the ice increased drastically around -10 °C. Then the origin of the adhesion force changed from the interaction between the liquid-like layer and polyethylene to the solid ice and polyethylene, and then the attractive interaction force became weak but the destruction force (shear force) of the interface material increased because the ice is harder than liquid-like layer (interfacial matter is changed from liquid-water to solid-ice). These effects seem to associate with increased brake effect at very low temperature.

Decreasing the μ value of high velocity at high temperature (Fig. 3) was not yet explained, but the strong evidence for the dry process indicated that the trace of sliding was observed on the flattened surface on snow truck (Fig. 7). It is reasonable to assume that the ice became soft near melting point then the shearing force also would be weak.

Figure 7. A micrograph with polarized light of a highly polished area of snow surface.

4. Conclusion

For the elucidation of the mechanism of friction between snow and ski, the kinetic frictional coefficient (μ) of a short slider at low gliding velocity (0.001~1 m/s) were observed near the melting point of snow (-20 ~ -1 °C). As the experimental results, the μ values were dependent on sliding velocity and snow temperature. The μ value of fast glides (higher than 0.5 m/s) near the melting point was decreased but an obvious pressure effect was not observed. It is unlikely that the melt-water is generated at around 1 m/s for any temperature for the model ski. The smallest μ value was observed at around -10 °C and slower than 0.01 m/s. It is interesting that the smallest μ value without melt-water is similar to the value of the best condition of usual skiing, which explained by the melt-water lubrication hypothesis. It is reasonable that as the liquid-like layer becomes thin accompanied with the decrease of temperature, the adhesion force also becomes small.

On the other hand, the μ value was increased with decreasing the temperature from -10 °C. This is explained by the shearing force of ice becoming large accompanied by the hardness of the ice increasing drastically around -10 °C.

The static frictional coefficient was dependent on the logarithm of the contact time. The results indicate that the real contact area decreased with increasing velocity. Further, the coefficient increased with increasing temperature for smaller grain snow. So the adhesion did not affect the sliding at high velocity and for the bigger grain size. It means the real contact area changed only slightly.

The sliding resistance of the model ski is consistent with the shearing force of the adhesion between snow and the polyethylene sole.

5. Acknowledgments

We are grateful to the members of the ski science association of UEC for the discussions that have contributed greatly to the investigation of this field. We also thank undergraduate course students of UEC for their assistance in our experiments. This work has been financially supported in part by Grant-in-Aids for Scientific Research Nos. 16500395 and 19500528 and the grants of Sugadaira Space Radio Observatory, the University of Electro-Communications, Nos. 1996-1, 1997-1 and 1998-1.

Journal of Physics: Conference Series **258** (2010) 012007 doi:10.1088/1742-6596/258/1/012007

6. References

[1] Saitou K 1962 *Surface technological compendium* ed Youkendou 70
[2] Tabor D 1961 *Adhesion* ed D D Eley Oxford Univ. Press 115
[3] Tusima K 1977 *J. Glaciology* **19** 225
[4] Bowden F P and Hughes T P 1939 *Proc. R. Soc. Lond. Math. Phys. Sci.* A **172** 280
[5] Ambach W and Mayr B 1981 *Cold Reg. Sci. Technol.* **5** 59
[6] Glenne B 1987 *J. Tribol.* **109** 614
[7] Tusima K and Yoshida Z 1969 *Low Temperature Sci.* A **27** 17
[8] Colbeck S C, Najarian L and Smith H B 1997 *Am. J. Phys.* **65** 488
[9] Colbeck S C and Warren G C 1991 *J. Glaciology.* **37** 228
[10] Buhl D, Fauve M and Rhyner H 2001 *Cold Reg. Sci. Technol.* **33** 133
[11] Colbeck S C1992 *CRREL Monograph* 92-2 Hanover NH
[12] Hamalainen T and Spring E 1986 *Comment. Phys.-Math.* **76** 1
[13] Dieterich J H and Kilgore B D 1994 *Pure and Appl. Geophys.* 143, 283
[14] Dieterich J H and Kilgore B D 1996 *Tectonophysis* 256, 219
[15] Hidaka Y, Kai S and Matukawa H 2009 *Nonlinear Sci. Ser.* 4 ed Nakamura K (Tokyo) Baihukan P174
[16] Bowden F P 1964 *New Scientist* **376** 275
[17] Barnes P, Tabor D and Walker JCFT 1971 *Proc. Roy. Soc. Lond.* A **324** 127
[18] Rosenberg R 2005 *Phys. Today* December 50

Journal of Physics: Conference Series **258** (2010) 012008 doi:10.1088/1742-6596/258/1/012008

Dual characterization of boundary friction thanks to the harmonic tribometer:
Identification of viscous and solid friction contributions

M Belin[1], M Kakizawa[2], E Rigaud[1], JM Martin[1]

[1] Ecole Centrale de Lyon, LTDS, CNRS, 36 av. Guy de Collongue, 69134 ECULLY cedex, France
[2] Shiseido Research Center, 2-2-1 Hayabuchi, Tsuzuki-ku, Yokohama, Kanagawa, 224-8558, Japan

E-mail: michel.belin@ec-lyon.fr

Abstract. All the usual characterization of an interface submitted to shear is generally achieved thanks to an experiment using a pair of two loaded solids in continuous sliding. So, the friction coefficient value can be determined. But, only this friction value is quite poor information and is not able to reveal the complexity of the interfacial processes occurring in the sliding contact. A novel method has been developed at LTDS, allowing us to identify both the velocity-dependent and the solid friction contributions. It is based on a contact submitted to constant normal load, one solid being supported by a 1D mechanical oscillator, under the form of an elastic bi-blade, able to return to its equilibrium position through damped oscillations. The way the amplitude is decaying during this elastic recovery is recorded. A mechanical model has been developed, in order to fit experimental data, allowing us to identify two different contributions: the velocity-dependent contribution (typically viscous damping), and the solid-like contribution (typically solid friction). This technique has been applied here to a lubricated sphere-on-plane contact. The contacting surfaces are made of AISI 52100 polished steel. Two model products are tested: pure glycerol and 1,3-buthylene glycol. First results are presented, showing the capabilities of this technique for giving a new insight on the mixed lubrication regime description.

1. Introduction
The control of friction has been identified as a challenging topic for energy saving. Especially, reducing friction is a goal for many systems involved in transportation: engines, gears and transmissions. A good knowledge of phenomena occurring during friction is then needed. From a classical approach, friction is measured thanks to tribometers, in which two solid surfaces are rubbed together in given loaded conditions. The lateral force is measured and then the friction coefficient is simply calculated by dividing the tangential force by the normal one. This classical approach is extensively used in the different fields of tribology, for both fluid and solid lubrication. It gives overall information relative to the interface under sliding but does not bring any detail concerning the intimate processes involved into the velocity accommodation. In addition, different tribological systems have been identified with low friction. Recent developments on so-called "Superlubricity" have been shown [1]. In such cases, the value of classical friction coefficient can be as low as 10^{-3} to 10^{-4} or even less. Some theoretical results have even predicted a total vanishing of friction force in particular conditions

[2]. In such case, the determination of friction coefficient value with a sufficient precision is very difficult to obtain, due to limitations of the force transducers [3,4].

For these two reasons, a new technique has been developed. This paper presents some preliminary results obtained thanks to an original alternative way that allows to identify linear velocity-dependent and velocity-independent friction contributions, without measuring any tangential force. It is based on the development of a new instrument, and the analysis of the dynamic free response of a single degree-of-freedom mechanical oscillator.

2. Principle of the technique developed

2.1. Basic principle and apparatus

The basic principle of this method is based on the analysis of the dynamic free response of a single degree-of-freedom mechanical oscillator, to which a contact with friction is attached. The contact is a sphere-on-plane configuration. The initial situation to be considered is when the head is set out of equilibrium, with an initial deviation parallel to the sliding direction. When the head is released, relaxation process occurs, and the elastic energy accumulated in the elastic blades is progressively released. Position and speed of the head is then oscillating, and their amplitude is decreasing with time. The detailed analysis of the damped relaxation oscillations leads to qualitative and quantitative information on friction. This mechanical problem has been presented and solved in detail in literature [5,6]. We assume that friction at interface can have two origins: i/ a viscous-type component, depending on the sliding velocity, characterized by "**zeta**" value, ii/ and a solid-type friction component, characterized by "**mu**" value, Figures 1 and 2. A dedicated apparatus has been developed and presented previously in details [7].

2.2. Measurements and data processing

The relaxation process may have a total duration ranging from 20 sec (in case of low friction) down to less than 1 sec. Different signals are recorded during this phenomenon: i/ position and speed of the moving head, thanks to a laser velocimeter (POLYTEC OFV-5000), ii/ electrical contact resistance (ECR), thanks to a dedicated electronic device, described previously [7], iii/ the overall friction force, with a piezoelectric transducer, in order to have a qualitative approach of lateral force during relaxation. Identification is then processed on both position and speed, in order to get the final values of zeta and mu. Much care is needed in this data processing. Especially, It may happen that the first oscillations are deformed, due to some defect of positioning before relaxation. In such case, these oscillations are removed from the signal to be processed. No incidence on final values of zeta and mu is observed.

2.3. Precision and accuracy

The interest of this method is that identification process leads to a unique solution for both zeta and mu values. This technique gives good results for modeling both the number of oscillations and the amplitude decay evolution of position and speed of the moving head. Therefore, this method can be considered as very robust. The lower limit for the number of oscillations to be processed is 5 to 6, in order to have sufficient resolution. We consider that zeta becomes negligible when its calculated value is lower than 0.0001. The smaller the damping is, the better is the resolution for mu and zeta.

3. Experimental and results

3.1. Samples

We present here the results obtained with two different lubricants: glycerol $C_3H_8O_3$ (GL), with a dynamic viscosity of 0.72 Pa.s, and 1,3-buthylene glycol (13BG), with a dynamic viscosity of 0.09 Pa.s, as measured is a cone-plate rheometer at 25°C (ANTON PAAR MCR-300), with shear rate ranging typically between 10 and 1000 s^{-1}. The contacting surfaces are a ball, 3.0 mm diameter and a flat, both made of 52100 AISI steel, polished by a series of diamond pastes: 3.0 then 1.0 μm, in order to obtain mirror-polish surfaces.

3.2. Natural self-damping of the system

Experiments without contact between the spherical pin and the flat have been performed, in order to identify dynamic and damping characteristics of the test apparatus by itself. Fig. 3 displays oscillations of the head with spherical pin in air. The number of oscillations observed before equilibrium is larger than 300 in this case, see Table 1. The dynamic response corresponds to purely velocity-dependent dissipation and equivalent viscous damping zeta, identified from the least mean square method. The zeta value is found in the millirange: zeta = 0.0013. Therefore, modeling the device by a single degree-of-freedom linear oscillator is validated. Apparatus dissipation can have different origins: inner damping inherent to the materials used, damping caused by micro-sliding in the connections between assembled mechanical parts, damping caused by air displacement and also the acoustic radiation of vibrating surfaces and vibrations transmitted to the flat. The relaxation behavior is linear as the amplitude of initial displacement remains small compared to the biblade length.

Table 1. Response of the steel-steel contact in different conditions: free response (no contact, no meniscus), meniscus response (no contact) and loaded contact lubricated with glycerol (GL) and 1,3-buthylene glycol (13BG), for a fixed normal load 50 mN.

Normal load (mN)	Approx. number of oscillations	zeta	mu	ECR log(Ohm)
No lubricant				
free relaxation	>> 300	0.0013	/	/
GL lubricant				
0 (meniscus only)	ab. 100	0.0108	/	/
50	16	0.0186	0.037	>4, stable
50 (repeated)	16	0.0185	0.038	>4, stable
13BG lubricant				
0 (meniscus only)	ab. 300	0.0026	/	/
50	12	< 0.0001	0.179	<1, fluctuations
50 (repeated)	11	< 0.0001	0.192	<1, fluctuations

3.3. Effect of lubricant meniscus

In a lubricated contact, a meniscus of fluid is surrounding the real contact zone. The effect of meniscus is now under consideration. We have performed systematic experiments with a given quantity of lubricant, here 50 μl. As an example, we report the dynamic response of a sphere facing a plane sample, non-contacting and fed by GL lubricant. So a fluid meniscus is present in the gap (50 μm thickness) between solids. The time response is presented in Fig. 4. The number of oscillations before equilibrium is found close to 100, so much less than without lubricant, as expected. Identification of the time response leads to a centirange value: zeta = 0.0108. This value is one order of magnitude larger than without lubricant surrounding the contact. Therefore, we get evidence for the effect of the fluid meniscus in this technique. The same type of experiment is run for 13BG lubricant, having a lower viscosity. We can check that in this case, the damping is larger than without product, but much lower that in the more viscous case. Identification leads to a millirange zeta value: 0.0026.

From this section, we can understand that both the equipment itself and the fluid meniscus bring a viscous-type contribution to the harmonic relaxation of the moving head, after releasing out from its out-of equilibrium initial position.

3.4. Response of the lubricated contact with normal load

We have performed experiments with lubricants at a given normal load of 50 mN. The corresponding maximum contact pressure is 245 MPa and the corresponding Hertzian diameter of 20 μm. The results are presented in Table 1.

3.4.1. Glycerol (GL)

For glycerol, we can observe that the global damping of the lubricated loaded contact is larger than that for the non-contact situation (meniscus only), as expected, Figure 5.a. Damping is shown to increase when the normal load is applied, leading to a decrease of the number of oscillations before equilibrium. Identification of the position response leads to a centirange value of the solid-type friction: 0.037 and a millirange value of the viscous-type friction 0.0186. The results show that both the relaxation overall duration and the decay law of the oscillating amplitude is well fitted. We periodically checked the repeatability of the results. As an example, we show in Table 1 the results obtained in two different experiments for the normal load of 50 mN. The values of zeta and mu are found to be very close. ECR evolution shows that the measured value is set to more than 10^7 ohms Figure 5.b, showing that the contact is in full film conditions. For the last three oscillations, these conditions are changing and some metal junctions are appearing, giving some bumps on the ECR signal. The final value of ECR correspond to metal-metal contact.

3.4.2. Buthylene glycol (13BG)

We have also performed experiments with the lower viscous product, 1,3-buthylene glycol (13BG), Figure 6.a. We can observe that the damping is higher than in the previous case, leading a significant decrease of the oscillations. By identification, the two components are calculated. The solid-type contribution one found in the decirange: 0.179 and the viscous-type component is found negligible (<0.0001). Repeatability of results is shown in table 1. It seems that in mixed regime, dissipation can be influenced by some roughness effect. ECR signal shows clearly that mixed lubrication is occurring in the contact, Figure 6.b.

4. Discussion

4.1. Fully fluid lubrication regime (GL case)

The experiments achieved with GL lubricant show that the contact is operating with a fully fluid film in the contact, thanks to ECR measurement. ECR has a high value -close to 10^5 ohms- and stable with time. For this reason, in-situ electrical measurement is very useful and no doubt is possible on this point, Figure 5.b. It means that the contact is fed by a full film, whatever the speed value. In these experiments, the maximum sliding speed is 70 mm/s. In the classical EHL approach, the contact is assumed to operate in steady state conditions, with a constant value of sliding and rolling speed conditions. Then the fluid film thickness is strictly depending on the speed value. Therefore, if we consider that the film in the present contact is set due to combined effect of sliding speed, viscosity and piezo-viscous effect [8]. In this technique, the sliding speed at the contact is essentially varying, in a damped sine wave motion. Especially, when speed is passing by the null value (twice by each oscillation), a film of lubricant is remaining in between the two contacting bodies. In GL lubricated case, we can observe from measurement that ECR value is stable at the maximum value This phenomenon can be considered as an effect of squeeze film, as already described in literature [9]. This phenomenon can be quite important for applications, because we show that dynamic lubrication can be achieved even when the speed is zero.

When the relaxation comes to the end, in the last two oscillations, we observe that ECR value decreases drastically, showing that the complete fluid film is no longer operating between the surfaces. When checking the lateral force that is measured during relaxation, Figure 5.c, we can have a qualitative evolution of the total frictional force at the contact. The signal has not been calibrated here, so the evolution is only qualitative. We can observe that the extrema of the modulating lateral force Ft are slightly decreasing (viscous contribution decreasing due to the progressive velocity reduction), then the shape of oscillations becomes more square like. During the last oscillation, we notice that the friction signal increases, just before stop. This can be related to the end of the full fluid film regime. stable, unless the speed reaches a low value. In this case, ECR is decreased. The lubrication regime is changed: from full film to mixed or boundary lubrication. This observation validates the mechanical model, in which we consider that the time response can be identified by a unique combination of both viscous-type and solid-type friction.

When considering the results of identification for the full film lubrication situation (GL), identification of the time response leads to a centirange value of both zeta and mu. It means that the shearing cannot be considered only as viscous-type flow. This unexpected result brings a new description of sliding contacts. We can assume that it is related to the physical change of the lubricant under high pressure contact, when the lubricant is expected to change from viscous to solid-like material, as described extensively in EHL literature [8].

4.2. Mixed/boundary lubrication regime (13BG case)

The results obtained with the 13BG show that the contact is now operating in mixed or boundary lubrication regime. On ECR signal, we can observe low values, with many fluctuations, showing that the contact is mainly supported by solid-solid asperities in contact, without a continuous fluid film existing in the interface. In these conditions, the solid-type friction is found in the decirange, typically less than 0.2 and the viscous-type component is found very low, less than 0.0001, Figure 6.c. This value can be considered negligible compared to decirange solid-type friction., This case is found to be very different from the previous, obtained with GL lubricant, more viscous. We can deduce that this technique can be beneficially applied to the study of mixed lubricated contacts, in order to learn more about the limits between boundary and mixed regime. This topic is currently widely studied in literature [ref], and many authors are presenting data on the effect of roughness, texturing of surfaces, lambda ratio on the shape of the Stribeck curve. The practical applications of such developments are important in terms of minimizing friction, increasing durability of sliding contacts [10,11].

4.3. Additional friction measurements during relaxation

Basically, the new technique presented in this paper does not need any force measurement. The analysis is based on the study of the decay curves of both position and speed of the relaxing head with time. Then, the mechanical model allows us to identify viscous damping component and friction component generated by interfacial shearing. But in addition, a piezoelectric force transducer has been introduced in the sample holder, so that the measurement of the variation of tangential force is made possible. This measurement is still difficult, due to the low level of forces under consideration. The system has not been yet calibrated precisely and it gives only qualitative information, as presented in Figure 5.c and Figure 6.c. From variations of amplitude, phase and shape of these tangential force signals, we can think that precious information will be extracted in the future. This work is currently under progress. So, the complete experimental technique presented here is expected to bring a better understanding of complex friction phenomena, like pure viscous shearing, slip at wall interface [12], shearing of a solid-like interfacial media.

5. Conclusions

This paper presents a new method for evaluation of friction in a sliding contact. It is based on an experimental arrangement in which the sliding contact under load is supported by an elastic system. When the head is set out of equilibrium and relaxing, the study of damped oscillations process is performed. A mechanical model based a 1D oscillator is used, and the main assumption consists on the possible contributions of both a viscous-type friction and a solid-like friction. Position and speed of the moving head is achieved by a laser velocimeter. In addition, electrical contact measurement and lateral force signal is recorded.

From the time response, we have been able to identify two components of the frictional dissipation: viscous-type friction and solid-like friction. The machine itself has been qualified. The results show that is behaves like a pure viscous damping system (zeta= 0.0013). The technique has been applied to a steel on steel contact lubricated by 2 different products: pure glycerol (GL) and 1,3-buthylene glycol (13BG). For GL, the time response gives evidence for a fully fluid contact, thanks to ECR. Zeta and mu values are determined and both are found in the centirange. For 13BG: evidence is given for a boundary lubrication regime. The friction is mainly solid-type: mu in the decirange value and the viscous-type damping is found negligible in this case. The capabilities of this new technique are expected to bring a new insight in friction measurement and characterization.

The further steps in developments could concern the following items:

i/ Some improvements of the experimental rig, especially the calibration of lateral force signal, when measurable,

i/ Application of this technique to different tribological systems operating in mixed and boundary lubrication regimes: effect of surface effective additives like adsorbed layers for example, effect of surface patterning for the case of controlled roughness. This technique could bring some new description of friction dissipation, for example the detection of viscous shearing, slip at solid surface, internal shearing, *etc.*

ii/ Application to superlow friction systems, when friction measurement is made difficult due to the low values of lateral force to detect. For that purpose, different tribological systems could be under consideration, both for fluid lubrication and solid lubricants, like MoS_2, DLC, ta-C for example. Let us remember that with this method, lower is friction -and damping, better is precision;

iii/ We can also propose that such an approach can be beneficially applied to a tip-probe based contact, like AFM-type equipments. In such case, the single asperity contact would make possible to elucidate the intimate processes occurring during friction.

References

[1] Martin JM, Donnet C, Le Mogne T, and Epicier T. 1993 Phys. Rev. B 1993 **48** (14)
[2] Hirano S. 1990 Phys. Rev. B **41** 11837-11851
[3] Schmitz T, Action J, Ziegert J and Sawyer G. 2005 Journ. of Tribology **127** (4) 673-8
[4] Burris D, Sawyer G. 2009 Tribology Letters **35** (1) 17-23
[5] Feeny B, Liang J. 1996 Journal of Sound and Vibration **195** 149-154
[6] Liang J, Feeny 1998 Phys. Rev. B. **16** (4) 337-47
[7] Rigaud E, Perret-Liaudet J, Belin M, Joly-Pottuz L and Martin JM 2010 Trib. Int. **43** 1-2 320-329
[8] Vergne P, Superlubricity 2007 Ed. Erdemir A and Martin JM, Elsevier, 427-443
[9] Lin JR 2000 Computers and Structures **75** 73-80
[10] Kovalchenko A, Ajayi O, Erdemir A, Fenske G and Etsion I 2005 Trib. Int. **38** 219-225
[11] Costa HL., Hutchings IM. 2007 Tribology Int. **40** 1227-1238
[12] Zhu Y and Granick S, Superlubricity 2007 Ed. Erdemir A and Martin JM, Elsevier 103-115

Figure 1. *Schematic arrangement for the measurement with the harmonic tribometer. The contact is a sphere-on-plane configuration. The sphere is attached to the head, fixed at the end of an elastic arm that can oscillate. The speed and position of the head is measured by a laser velocimeter. In addition, electrical contact resistance ECR between sphere and plane and lateral signal force on the plane sample are measured during friction. Below, the schematic of the mechanical analysis.*

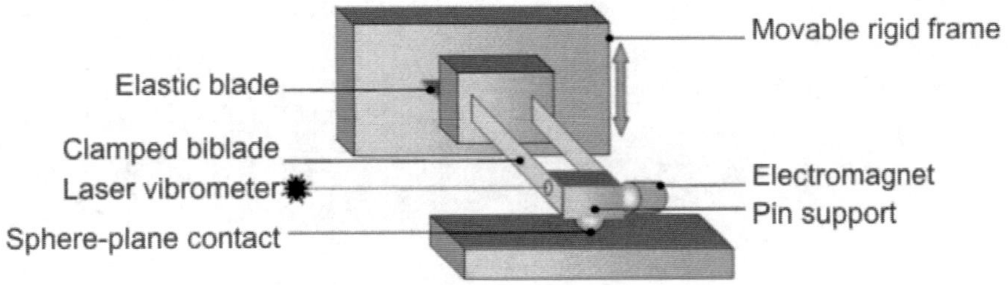

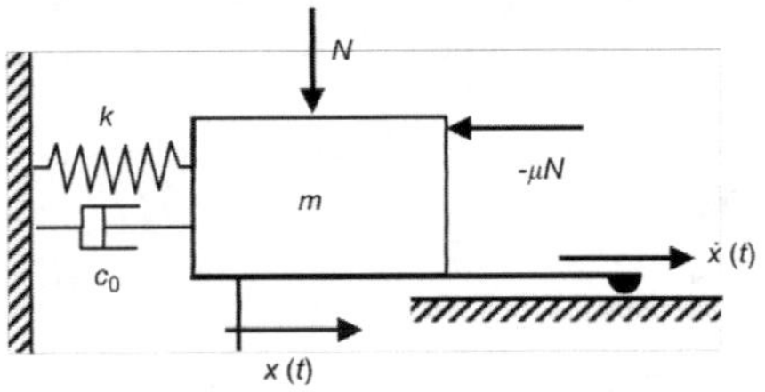

Figure 2. *Principle of the mechanical model. Friction is assumed to have two different contributions: i/ a viscous-type one, velocity-dependent and characterized by zeta value. The amplitude of oscillations during relaxation is expected to be exponential, ii/ a solid-type one, characterized by mu value. In this case, the decay of oscillations is expected to be linear.*

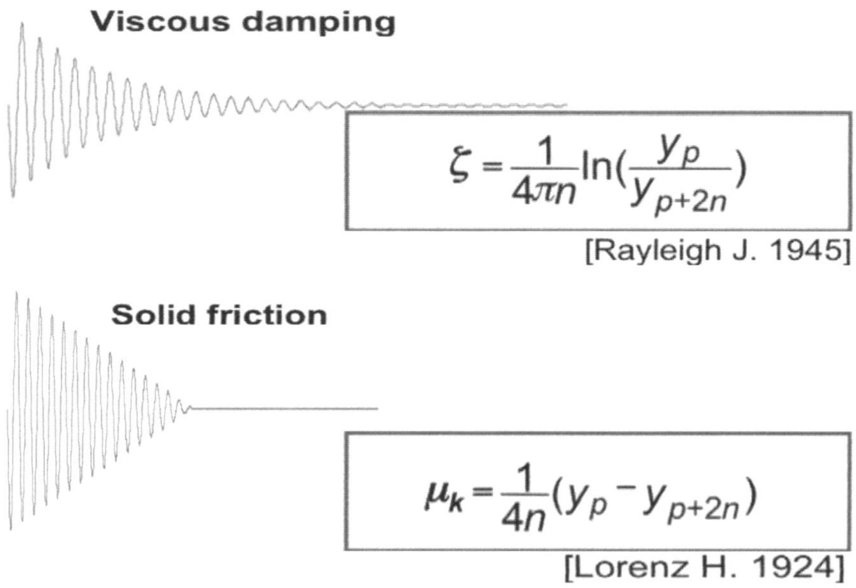

Journal of Physics: Conference Series **258** (2010) 012008 doi:10.1088/1742-6596/258/1/012008

Table 1. *Response of the steel-steel contact in different conditions: free response (no contact, no meniscus), meniscus response (no contact) and loaded contact lubricated with glycerol (GL) and buthylene glycol (13BG), for a fixed normal load 50 mN.*

Normal load (mN)	Approx. number of oscillations	zeta	mu	ECR log(Ohm)
No lubricant				
free relaxation	>> 300	0.0013	/	/
GL lubricant				
0 (meniscus only)	ab. 100	0.0108	/	/
50	16	0.0186	0.037	>4, stable
50 (repeated)	16	0.0185	0.038	>4, stable
13BG lubricant				
0 (meniscus only)	ab. 300	0.0026	/	/
50	12	< 0.0001	0.179	<1, fluctuations
50 (repeated)	11	< 0.0001	0.192	<1, fluctuations

Figure 3. *Time response of the relaxing moving solid, after initial displacement. The evolution of position is plotted. We can observe the characteristic decay of amplitude over 10 sec. A detailed plot is given in the graph below, showing the oscillations, more than 300. The model is based on a single component behavior: zeta= 0.0013.*

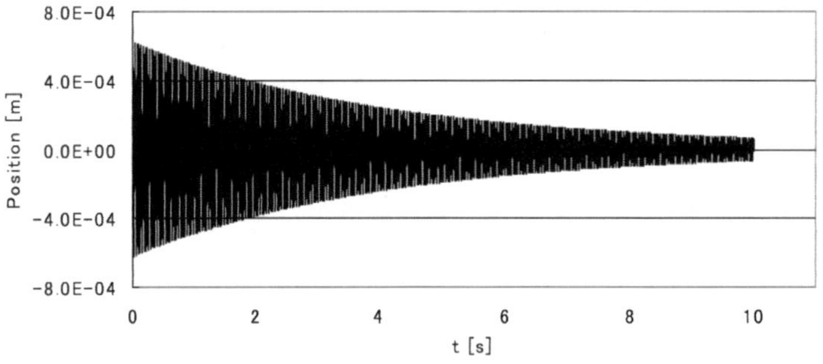

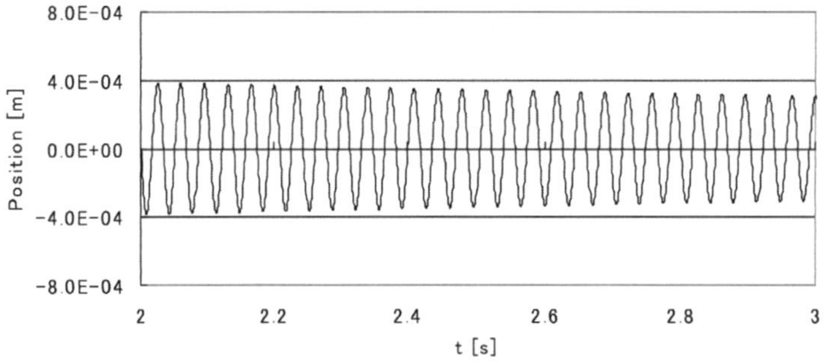

Figure 4. *Time response of the relaxing moving solid, fed by 50 μl of glycerol (GL), after initial displacement. The evolution of position is plotted. We can observe that the duration of relaxation process is shorter (about 3 sec.), showing that the damping is more intense. Identification leads to a single component behavior: zeta= 0.0108.*

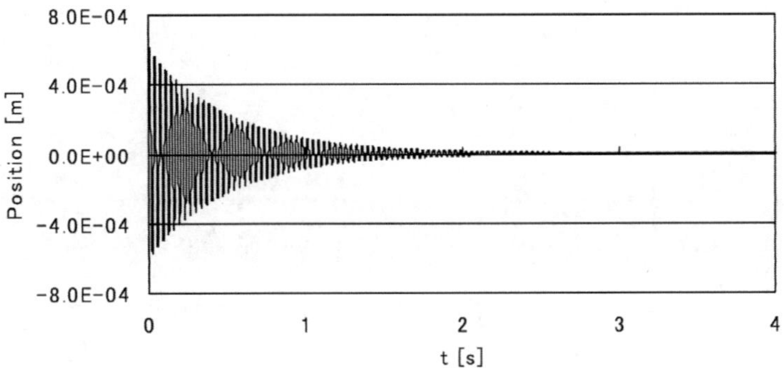

Figure 5. *Time response of a sphere-on-plane contact lubricated by pure glycerol: GL, Fn= 50 mN, temp= 24°C, AISI 52100 steel against itself.*
a/ Evolution of position: both experimental data for position (full line) and computed response (dotted line) are plotted. The model is based on a dual component behavior: zeta= 0.0186 and mu= 0.037.
b/ Evolution of electrical contact resistance ECR, showing the full fluid film lubrication regime
c/ Evolution of the lateral force signal, giving qualitative evolution of the friction force during relaxation.

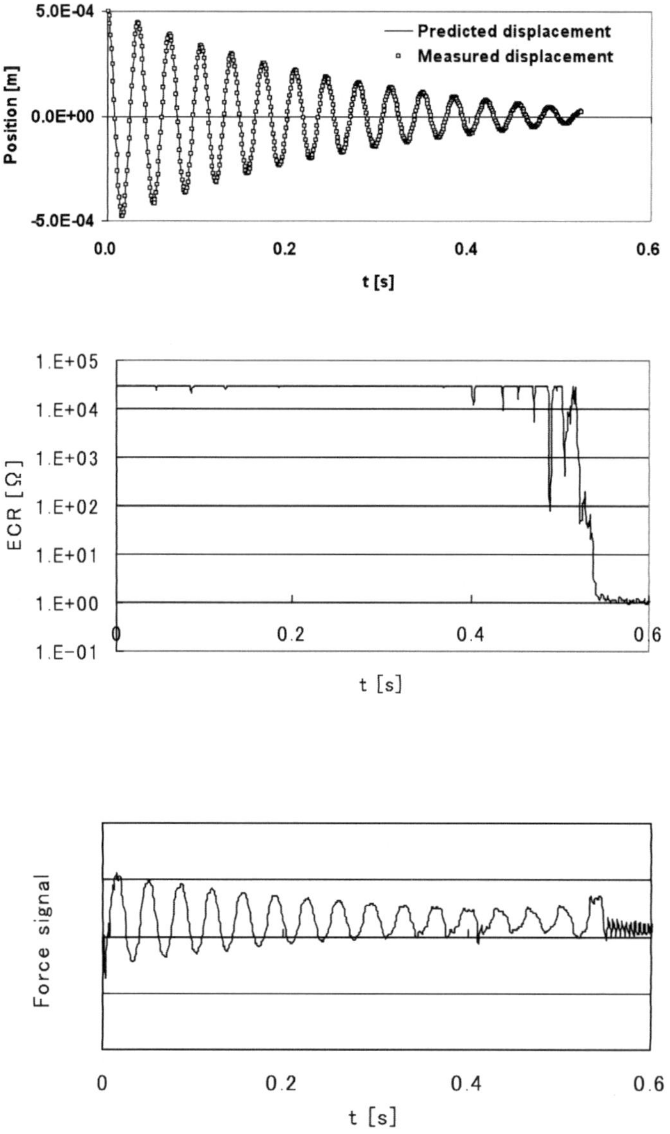

Figure 6. *Time response of a sphere-on-plane contact lubricated by buthylene glycol: 13BG, Fn= 50 mN, temp= 24°C, AISI 52100 steel against itself.*
a/ Evolution of position: both experimental data for position (full line) and computed response (dotted line) are plotted. The model is based on a dual component behavior: zeta< 0.0001 and mu= 0.179.
b/ Evolution of electrical contact resistance ECR
c/ Evolution of the total force signal, giving qualitative evolution of the friction force during relaxation.

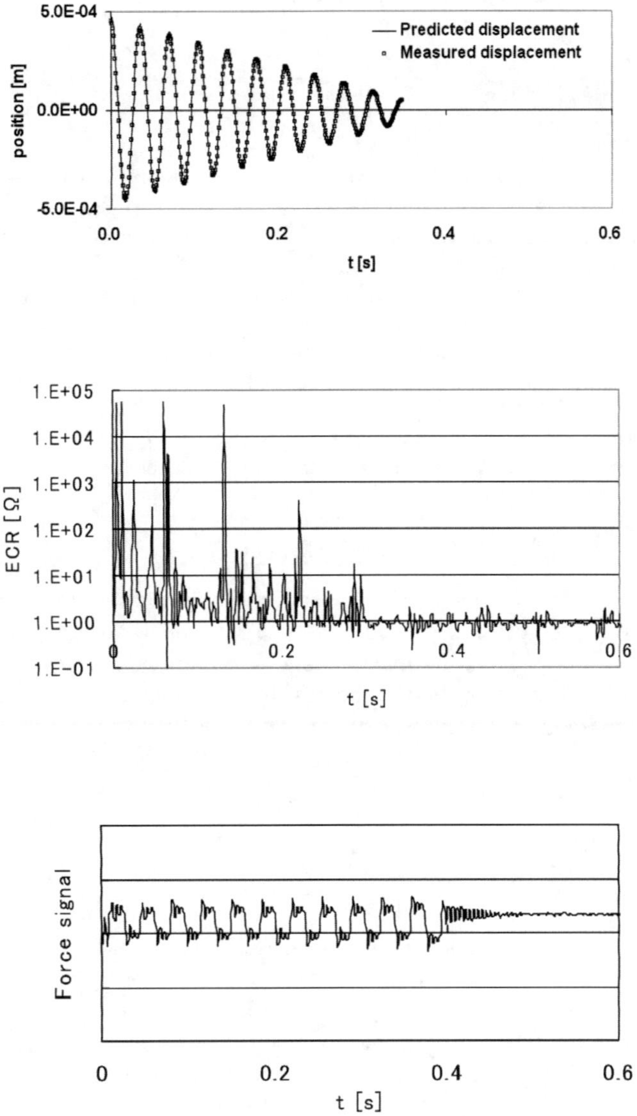

Journal of Physics: Conference Series **258** (2010) 012009 doi:10.1088/1742-6596/258/1/012009

Ultra Low Friction of DLC Coating with Lubricant

M Kano[1], K.Yoshida[2]
[1,2]Kanagawa Industrial Technology Center
705-1, Shimo-imaizumi, Ebina, Kanagawa
Japan

E-mail:.mkano@kanagawa-iri.go.jp

Abstract. The objective of this study was to find a trigger to make clear a mechanism of the ultra low friction by evaluating the friction property of DLC-DLC combination under lubrication with the simple fluid. The Pin-on-disc reciprocating and rotating sliding tests were conducted to evaluate the friction property. The super low friction property of pure sliding with the ta-C(T) pair coated by the filtered arc deposition process under oleic acid lubrication was found at the mixed lubrication condition. It was thought that the low share strength tribofilm composed of water and acid seemed to be formed on ta-C sliding interface. Additionally, the smooth sliding surface formed on ta-C(T) was seemed to be required to keep this tribofilm. Then, the super low friction was thought to be obtained by this superlubrication condition. Although the accurate and direct experimental data must be required to make clear this super low friction mechanism, the advanced effect obtained by the simple material combination is expected to be applied on the large industrial fields in near future.

1. Introduction

Reduction of the mechanical friction especially for automotive engine has been strongly required in recent years for improving fuel economy. Reducing the friction between the cam and follower is effective in this regard in the low engine speed range. A diamond-like carbon (DLC)-coated cam follower of the bucket type has been applied to the gasoline engines to reduce engine friction[1]. This low friction property is obtained by the hydrogen free DLC, ta-C lubricated with the ester containing oil[2]. The objective of this study was to find a trigger to make clear a mechanism of the ultra low friction by evaluating the friction property of DLC-DLC combination under lubrication with the simple fluid.

2. Test Method

The Pin-on-disc reciprocating and rotating sliding tests were conducted in the following manner. The pins, measuring 9 mm in diameter and 9 mm in length, were made of hardened bearing steel (AISI52100) and polished to a surface roughness below 0.05 micron. The disc measured 33 mm in diameter and 3 mm in thickness and was made of AISI52100 and polished to a surface roughness below Ra 0.05 micron. Three kinds of DLC were coated on those discs and pins. Those were a-C:H coating by Plasma Enhanced CVD with 1micron thickness, ta-C coating by the multi-arc PVD method with 1micron thickness and ta-C(T) coating by the T-Filtered arc PVD method[3] with 0.3 micron thickness. The droplets of the ta-C coating were removed by polishing.

---

[1] M.Kano, Kanagawa Industrial Technology Center.

The pin was secured to prevent it from rotating and was pressed against the reciprocating or rotating disc as shown in Fig.1. Contact at the sliding interfaces was in the shape of line under high Hertzian pressure of 70 MPa. The maximum sliding speed was 0.05m/s and the stroke was 20mm for the reciprocating test. The sliding speed was 0.05m/s for the rotating test. Lubrication with oleic acid and oleyl alcohol was provided by wetting on disc with several droplets (0.01ml) at ambient temperature.

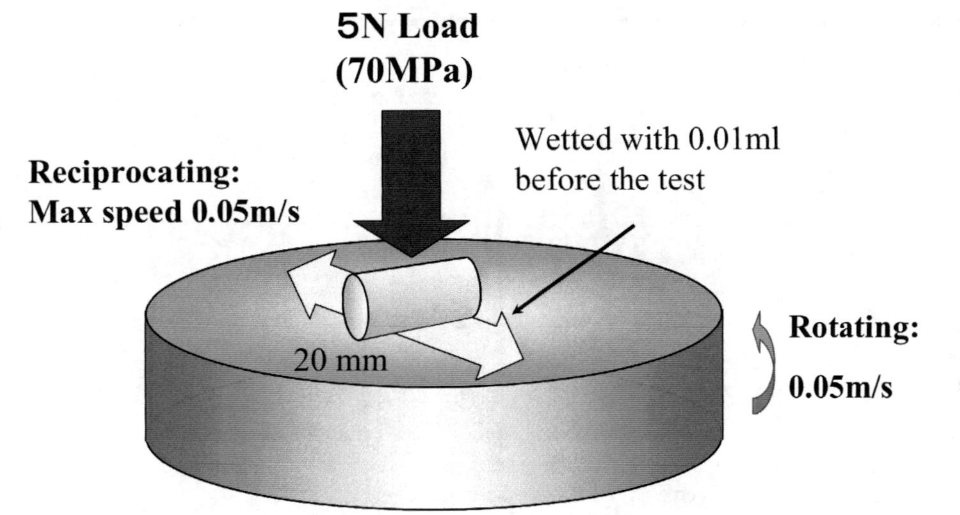

Figure 1. Pin on disc sliding test

3. Results and Discussions

The friction properties of DLC-DLC and the steel-steel pair after sliding for 30 minutes without lubrication (ambient air), with oleic acid lubrication and oleyl alcohol lubrication by the reciprocating test were compared in Fig.2. The friction coefficient of ta-C(T)-ta-C(T) pair under lubrication at 30 minutes was much lower than those of the other pairs, especially for the oleic acid lubrication. Additionally, the much different friction curves were observed for the sliding time from 0 to 30 minutes as shown in Fig.3. The friction coefficient of the steel-steel pair was very low level, 0.02 for 300 seconds and increased gradually to 0.085 finally. That of the a-C:H-a-C:H pair decreased from 0.08 to 0.045 for 1200 seconds, then kept the stable level. That of the ta-C-ta-C pair decreased continuously from high level of 0.09 to 0.045. On the other hand, that of the ta-C(T)-ta-C(T) pair decreased from very low level of 0.02 to the super low friction regime below 0.01 in very short time and kept very low level of 0.015 for 1800 seconds. Then, the original friction curves for the a-C:H-a-C:H and the ta-C-ta-C pair were shown in Fig.4. The friction coefficient for the ta-C-ta-C pair decreased to the super low friction in only 30 seconds from the starting point. On the other hand, that of the a-C:H-a-C:H pair did not decrease largely in 45 seconds.

The condition of the oleic acid fluid on the disc was observed as shown in Fig.5. The shape of the oleic acid for the ta-C(T) pair was enlarged in the whole sliding track after the sliding test compared with that of the a-C:H pair. This result suggests that the wettability for the ta-C(T) pair to the fluid must be changed better than that of the a-C:H pair and the tribochemical reaction between ta-C(T) and oleic acid might be generated during sliding.

The friction properties of DLC-DLC under oleic acid lubrication were also evaluated by the rotating sliding test. The friction coefficient for the ta-C(T) pair decreased rapidly to the super low friction regime about 0.005 as shown in Fig.6. Those of the ta-C pair and a-C:H pair decreased gradually to 0.025 level. The shapes of these friction curves except the steel pair were much similar to those for the reciprocating test in Fig.3. On the other hand, that of the steel/steel did not decrease much compared with that for the reciprocating test.

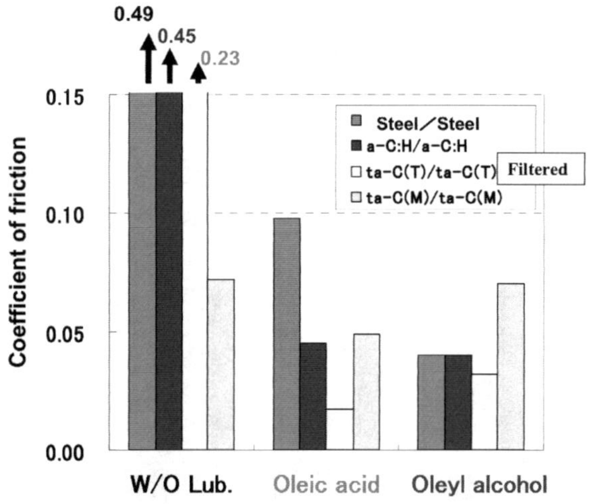

Figure 2. Friction coefficient of DLC-DLC pair

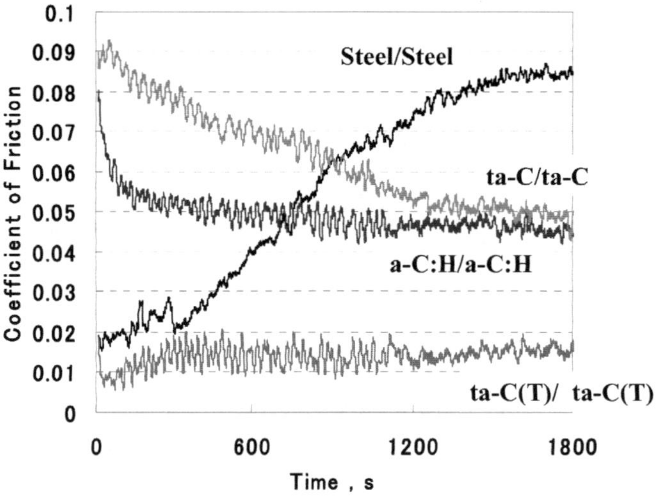

Figure 3. Different friction curves for DLCs

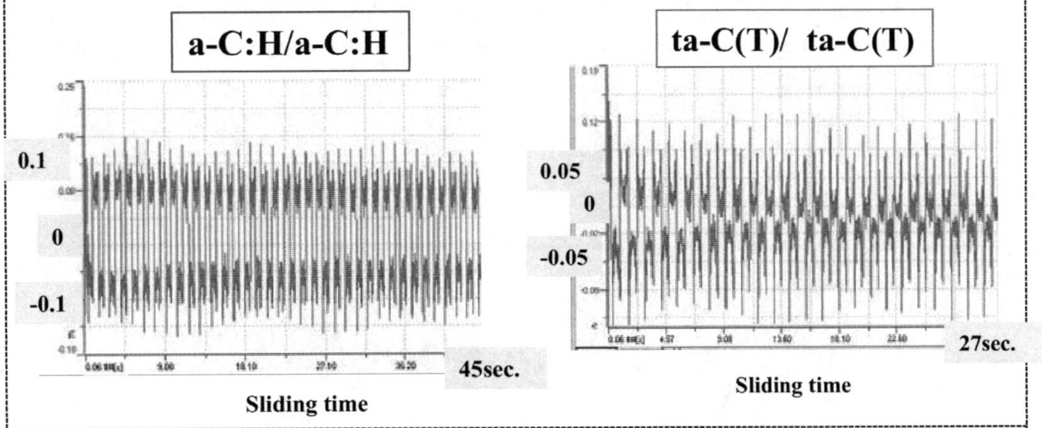

Figure 4. Friction curves from starting point

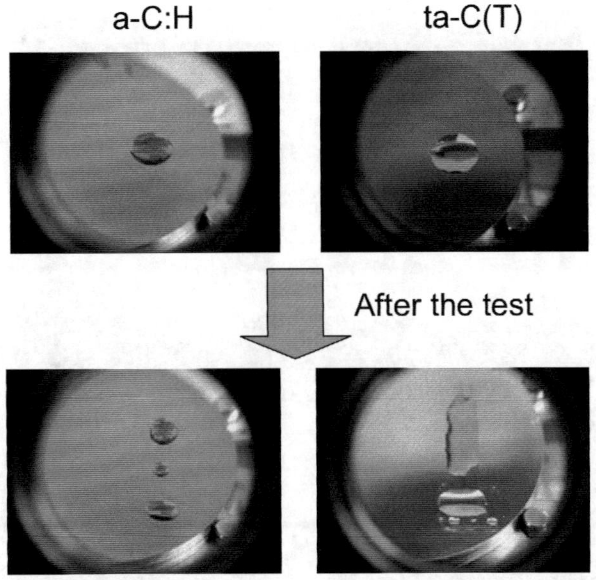

Figure 5. Condition of oleic acid on sliding track

The wear condition on the sliding track was observed by the scanning electron microscopy. Although a few numbers of shallow and circular grooves without the clear delamination were observed on the discs, more numbers of deep and linear grooves were observed on the pin. The pin wear seemed to be severer than that of the disc because the contact area of the pin was always same during the test. The wear conditions on the pin for the a-C:H and ta-C(T) pin were shown in Fig.7. The dark gray colored area was the un-sliding area, the light gray area was the sliding scar and the white colored lines were deep grooves in the upper photograph (low magnification). The deep grooves were formed in the middle of the wear track. Judging from the high magnification photograph, the DLC layer seemed to be delaminated partially for the ta-C(T) pin but the surface roughness around the grooves kept very smooth. Additionally, the wear debris for the a-C:H and the ta-C(T) seemed to be very fine with submicron size and the granular shape. Therefore, it seemed to keep very smooth sliding surface around the grooves and keep low friction coefficient even if the shallow grooves were formed partially.

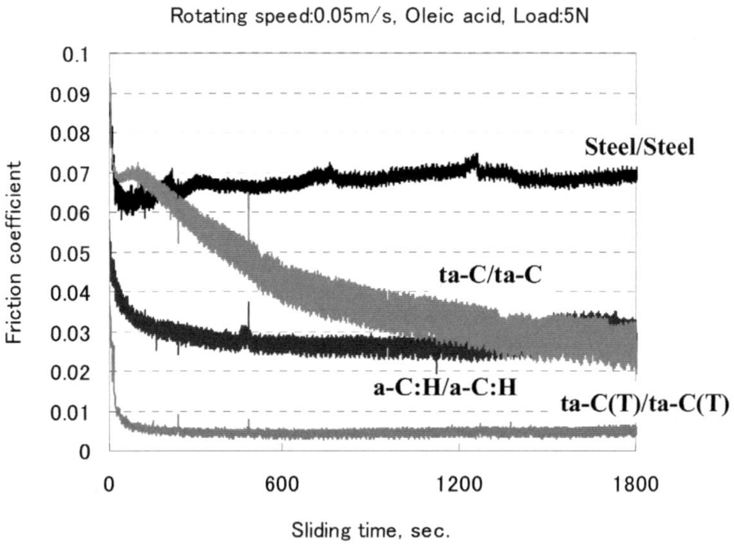

Figure 6. Friction coefficient of rotating sliding test

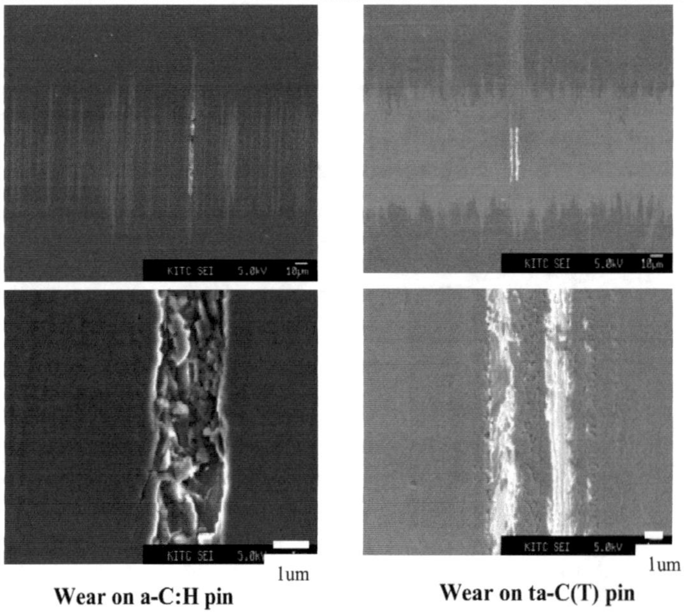

Figure 7. Wear on a-C:H and ta-C(T) pin after rotating test

For the ta-C which had rougher surface roughness with the droplets generated at the deposition process, the friction coefficient of the reciprocating test decreased from very high level, 0.09 to 0.05 continuously for 1800 seconds in Fig.3 and that of the rotating test decreased from 0.09 to 0.03 in Fig.6. Those friction coefficients seemed to be decreased more. Then, the reciprocating sliding test for longer time (2hours) lubricated with oleic acid was conducted for the ta-C pair and the ta-C(T) pair, as shown in Fig.8. Although the friction coefficient for the ta-C pair reached to lower level of 0.03 at 2 hour, the ta-C film on the pin had deep grooves and the substrate was appeared. Therefore, the friction coefficient might decrease more if the ta-C film would remain uniformly. The performance of reducing friction of ta-C seemed to be almost same as the ta-C(T). This ultra-low friction is thought to be generated by forming very thin and weak tribofilm.

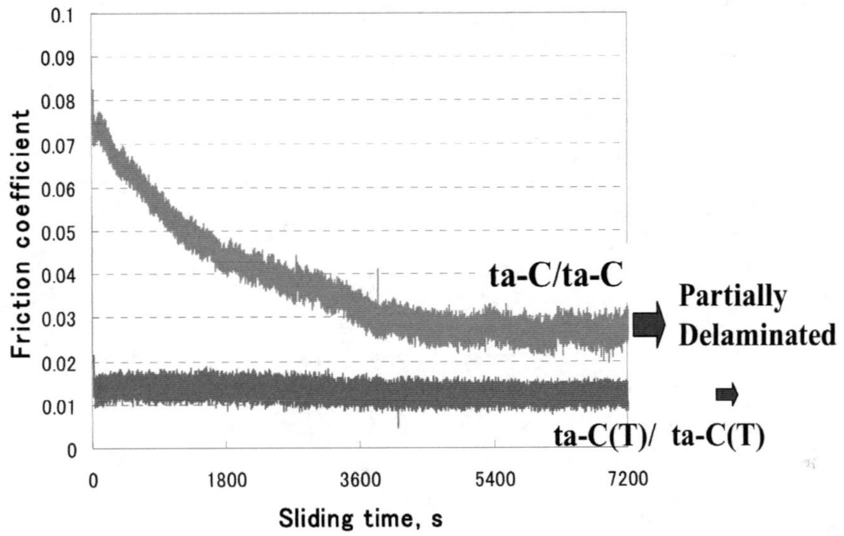

Figure 8. Reciprocating sliding test for long time

Then the relation between the friction coefficient and the combined surface roughness after the sliding tests was shown in Fig.9. As the combined surface roughness decreased, the friction coefficient decreased lineally except the steel pair. The initial surface roughness numbers for the steel pair, a-C:H pair and ta-C(T) pair were almost same low level, therefore the reason of this super low friction below 0.01 for the ta-C(T) was not thought that the sliding condition entered initially in the simple hydrodynamic lubrication. This ultra-low friction is thought to be generated by forming very thin and weak tribofilm at the mixed lubrication condition.

The friction curves of the DLC pair and the steel pair under oleyl alcohol lubrication by the reciprocating sliding test were shown in Fig.10. The friction coefficient for the ta-C pair and ta-C(T) pair did not decrease largely compared with those of the oleic acid lubrication in Fig.3. That for the a-C:H pair seemed to be almost same level as that for the oleic acid. Although that for the steel pair decreased to 0.03 level and kept low level, the tendency of increasing friction coefficient was observed. It seemed that the reducing friction performance for ta-C and ta-C(T) with oleyl alcohol was lower than that of oleic acid.

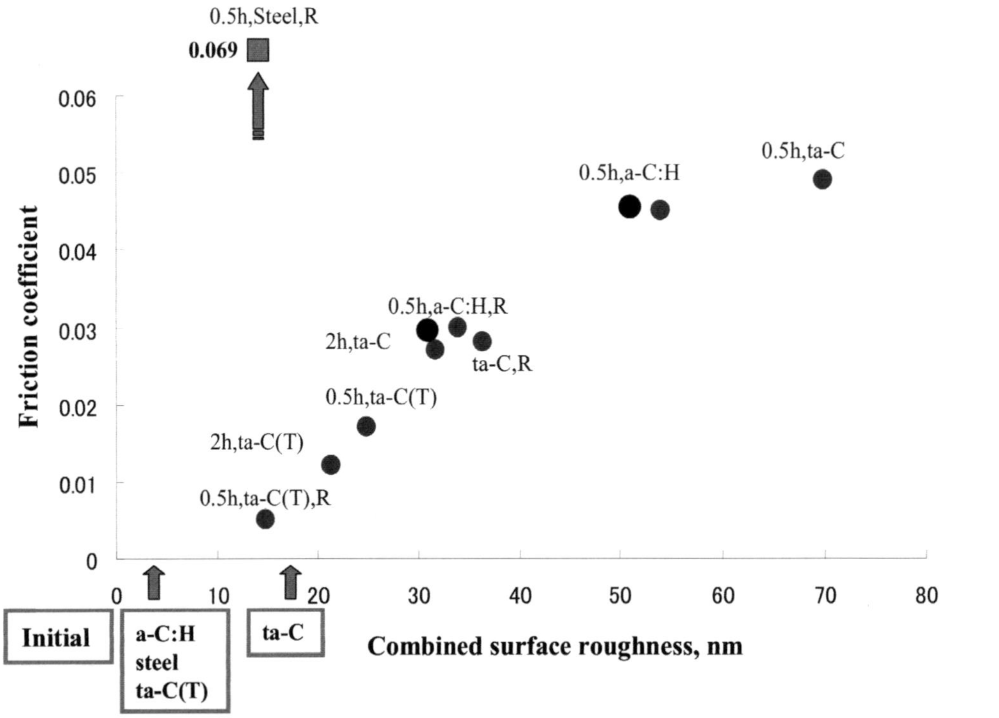

Figure 9. Relation between roughness and friction

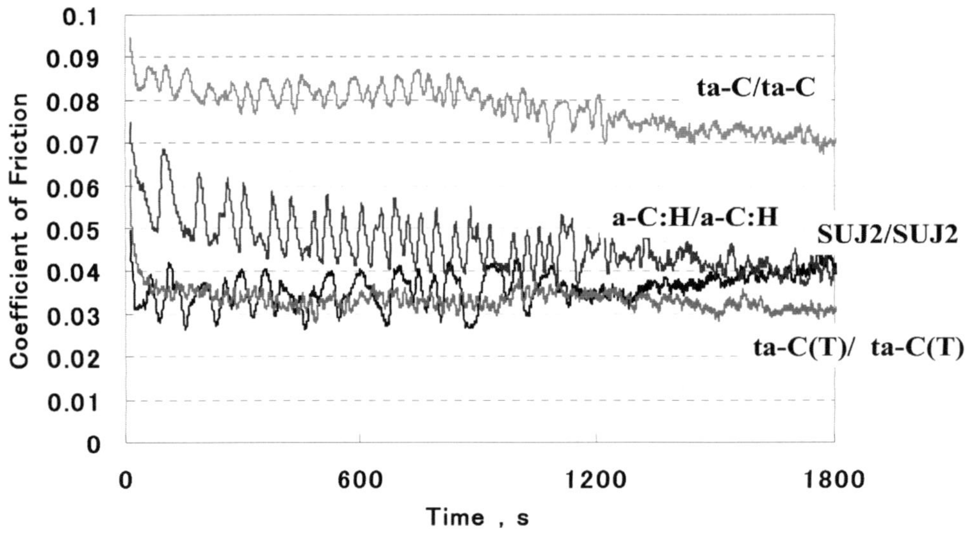

Figure 10. Friction curves for oleyl alcohol lubrication

J.M.Martin et al. proposed the possible mechanism of the super low friction was the forming water layer by the tribochemical reaction between ta-C and glycerol[4]. In this paper, the formation of water and acid was indicated as one possible chemical reaction for glycerol, as shown in Fig.11. For the oleic acid test, the water can be supplied from the ambient air. Therefore, the super low friction obtained by the ta-C(T) lubricated with oleic acid seemed to form the similar low share strength tribofilm at the sliding interface. Additionally, it was found that the surface roughness was required to be kept very smooth to maintain this tribofilm. Then, the super low friction was thought to be obtained by this superlubrication condition.

Dissociation of glycerol (P, T)

$$C_3H_8O_3 \Rightarrow C_2H_4O + CO + H_2O + H_2$$

acetaldehyde

Possibility of supercritical water

Other possible chemical Reactions:

$$C_3H_5(OH)_3 \rightarrow C_2H_5COOH + H_2O$$

$$2C_3H_5(OH)_3 \rightarrow 4HCOH + 2H_2O + C_2H_4$$

$$2C_3H_5(OH)_3 \rightarrow 2CH_3COOH + 2H_2O + C_2H_4$$

......

Figure 11. Possible tribochemical reaction for superlubrication

Although the accurate and direct experimental data must be required to make clear this super low friction mechanism, the advanced effect obtained by the simple material combination is expected to be applied on the large industrial fields in near future.

4. Conclusions

The super low friction property of pure sliding with the ta-C(T) pair coated by the filtered arc deposition process under oleic acid lubrication was found at the mixed lubrication condition. It was thought that the low share strength tribofilm composed of water and acid seemed to be formed on ta-C sliding interface. Additionally, the smooth sliding surface formed on ta-C(T) was seemed to be required to keep this tribofilm. Then, the super low friction was thought to be obtained by this superlubrication condition.

5. References

[1] S.Okuda, T.Dewa and T.Sagawa: JSAE20077162, SAE2007-01-1979
[2] M.Kano, Y.Yasuda, Y.Okamoto, Y.Mabuchi, T.Hanada, T.Ueno, J.Ye, S.Konishi, S.Takeshima, J.M.Martin, M.I.De Barros Bouuchet & T.Le Mogne: Tribology Letters, 18, 2(2005)245.
[3] H.Takigawa,K.Izumi, R.Miyamoto& T.Sakakibara: Surface Coating Technology, 163 (2003)368
[4] C.Matta, L.Joly-Pottuz, M.I.De Barros Bouchet, J.M.Martin, M.Kano, Q.Zhang and W.A.Goddard III: Physical Review B 78, 085436(2008)

Shear thinning behavior of monolayer liquid lubricant films measured by fiber wobbling method

Y Hamamoto [1,2], S Itoh [1], K Fukuzawa [1], H Zhang [3]
[1] Department of Micro-Nanosystems Engineering, Nagoya University, Furo-cho, Chikusa-ku, Nagoya, 464-8603, Japan
[2] Research Fellow of the Japan Society for the Promotion of Science 8 Ichiban-cho, Chiyoda-ku, Tokyo, 102-8472, Japan
[3] Department of Complex Systems Science, Nagoya University, Furo-cho, Chikusa-ku, Nagoya, 464-8601, Japan

E-mail: s_itoh@nuem.nagoya-u.ac.jp

Abstract. It is essential to clarify mechanical properties of monolayer lubricant films coated on magnetic disks under shearing motion for designing future hard disk drives with ultra-low flying height. Many of previous researchers reported that strong shear rate dependence of viscoelasticity was one of the typical phenomena observed with molecularly thin liquid films. However, it has not been clarified whether or not perfluoropolyether (PFPE) lubricant films, which are used for the head-disk interface (HDI) lubrication, show shear thinning behavior under actual HDI conditions. In this study, we used the fiber wobbling method that can achieve both highly-sensitive shear force measurement and precise gap control and measured shear rate dependence of viscoelastic properties of monolayer PFPE films coated on the magnetic disk. Our experimental results showed that shear thinning does occur at high shear rate ranged from 10^2 to 10^6 s^{-1}.

1. Introduction

To achieve further increase in recording density of hard disk drives (HDDs), the gap between the magnetic head and the disk must be reduced to less than 5 nm [1]. Taking into account this reduced gap, lubricant films used to lubricate head-disk interface (HDI) of HDDs must be a monolayer, with a thickness of around 1-2 nm. In addition, HDI must be designed to allow continuous or frequent contact between the head and the disk. In order to ensure durability and reliability of future HDDs, it is essential to characterize mechanical properties of monolayer lubricant films under shearing motion.

A surface force apparatus (SFA) is widely used to measure the viscoelasticity of molecularly thin liquid films. Ruths and Granick reported that the perfluoropolyether (PFPE) lubricants, which are used for HDI lubrication, showed shear thinning behavior under confined and sheared conditions [2]. On the other hand, no shear thinning behavior was observed with PFPE films by Bhushan's experiments using their original viscometer [3]. This discrepancy can be attributed to the differences in experimental conditions between the SFA method and Bhushan's method. Major differences are shear rate and materials of the substrate. In Bhushan's method, higher shear rate (10^4 – 10^7 s^{-1}) was achieved and the substrate was comparable to actual magnetic disks. However, the shearing gap was not well determined. In the SFA measurements, although the shearing gap was precisely determined, the shear rate was much lower (around 1 – 10 s^{-1}) and the mica was essential for the substrate. Since the

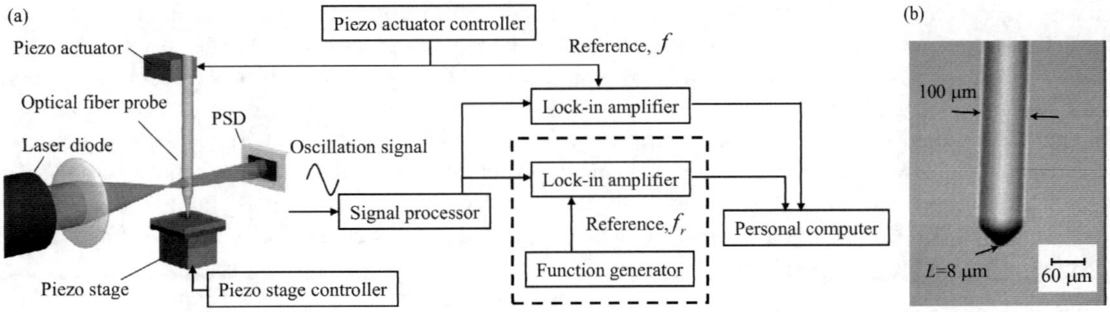

Figure 1. (a) Experimental setup for fiber wobbling method and (b) micrograph of optical-fiber probe.

shearing speeds dynamically change in the actual HDDs' operation, it is important for the lubrication design to clarify whether or not shear thinning occurs in HDI.

In our previous study, we originally developed a highly-sensitive shear force measurement method that utilized a ball-ended optical fiber as a shearing probe. We called this method the "fiber wobbling method (FWM)" [4, 5]. In FWM, the shearing probe was oscillated sinusoidally and the probe's tip sheared the sample to detect its dynamic viscoelasticity. We succeeded in measuring the viscoelastic properties of monolayer lubricant films coated on the magnetic disks [6]. There are many other methods using cantilevers placed orthogonal to the sample to achieve the sensitive detection of shear force [7-9]. One of the typical differences with our method was they used resonant oscillation of the probe to attain high sensitivity. In FWM, we can achieve high sensitivity without resonant oscillation. This means that we can choose any oscillation frequency to evaluated viscoelastic properties of samples. FWM also achieves a shear rate comparable to Bhushan's method. In this study, we used FWM to measure shear rate dependence of viscoelastic properties of monolayer lubricant films coated on the magnetic disk. Non-polar PFPE: Fomblin Z03 was used as the sample lubricant.

2. Measurement of viscoelastic properties of monolayer lubricant film by FWM

2.1. Principle of the fiber wobbling method (FWM)

The experimental setup for FWM is outlined in figure 1(a). The optical fiber probe positioned perpendicular to the sample was oscillated sinusoidally using a piezo actuator, and the lubricant-film sample on the substrate was sheared with the probe end. To measure the oscillation of the probe, the optical fiber probe was used as a cylindrical lens and the deflection of the probe was measured by detecting the displacement of the laser spot formed onto the position sensitive detector (PSD). The displacement signal from the PSD was synchronously detected with a lock-in amplifier referring to the forced oscillation signal for the piezo actuator. A laser Doppler velocimeter with detection limit of 1 nm (LV-1500, Ono Sokki) was used to calibrate the displacement signal. Our previous study showed that the detection limit of the displacement by using our technique was around 1-10 pm [10]. This detection limit corresponds to the shear-force detection limit of 1-100 pN considering that spring constant of the probe was usually around 1-10 N/m. The viscoelastic properties of the lubricant confined and sheared in the gap between the probe end and the substrate surface were obtained by measuring the change in probe amplitude and the phase shift.

In FWM, however, it is difficult to measure the shearing gap directly. Therefore, the gap width is determined by the displacement of the piezo stage from the point where solid contact occurred, which is a gap width of 0 nm. The accuracy with which the gap is controlled depends both on the resolution of the piezo stage and the sensitivity with which solid contact is detected. The resolution of the piezo stage is on the order of 0.1 nm. The point of solid contact is determined by using the following procedure. Both of the two shearing surfaces - the surface of the probe end and that of the substrate -

have surface roughness. Therefore, at the beginning of solid contact, asperity collisions occur and impulsive forces are applied to the probe. The impulsive forces can excite resonant oscillation on the probe. Thus, solid contact is detected by monitoring the excitation of the resonant-frequency component that is included in probe oscillation. The experimental setup that we used to detect solid contact is indicated in the dotted box in figure 1(a). To detect the resonant-frequency component in probe oscillation, we prepared another lock-in amplifier. A sinusoidal wave synchronized with the resonant frequency of the probe was used as the reference signal and it was input into the lock-in amplifier from the function generator. In our previous study, an accuracy of 0.2 nm was achieved in determining the solid-contact point [6].

There is a micrograph of the optical-fiber probe we used in this study in figure 1(b). The fiber diameter was about 100 μm, and the probe's end had a curvature radius of about 8 μm. The probe was 4.2 mm long. The probe's resonant frequency was 4945 Hz and its spring constant was 20 N/m. The surface roughness of the probe's end was measured by atomic force microscopy. The average roughness, root-mean-square roughness, and peak-to-valley roughness were 0.20 nm, 0.27 nm and 1.27 nm, respectively.

2.2. Calculation of viscoelastic properties

In viscoelastic measurement using FWM, the damping coefficient c_f and elastic coefficient k_f of the sheared samples are obtained as follows [4]:

$$c_f = \frac{a_0(k - m\omega^2)\sin\delta}{a\omega} \tag{1}$$

$$k_f = \frac{(a_0\cos\delta - a)(k - m\omega^2)}{a\omega} \tag{2}$$

where a and δ are the probe amplitude and phase shift measured at the shearing of samples, and a_0, ω, k, m correspond to the forced oscillation amplitude, the forced oscillation frequency, the spring constant of the probe, and the effective mass of the probe end, respectively. Dynamic viscoelasticity is often evaluated by the complex viscosity coefficients expressed as $\eta = \eta' - i\eta''$. Here, η' is the dynamic viscosity and η'' is the dynamic elasticity. We can relate η' and η'' to c_f and k_f as follows:

$$c_f = \eta'\Omega \tag{3}$$

$$k_f = \eta''\Omega \tag{4}$$

Here, Ω is the geometric parameter calculated from the shape of the sliding surface. If the sliding surfaces are flat plates and they move parallel to each other with the gap width of h, Ω is expressed as follows [11]:

$$\Omega = \frac{S}{h} \tag{5}$$

where S represents the contact area. Since the range of gap widths we focused on was sufficiently smaller than the curvature radius of the probe end, the geometric configuration of the sliding surfaces in our setup can be assumed to be two flat plates. However, it is difficult to determine an accurate value for S experimentally, because of the small curvature radius of the probe end, which was around 8 μm. Additionally, the dynamic deformation of the lubricant film expected during shearing motion makes it difficult to determine an accurate value. Therefore, we evaluated the viscoelasticity of the lubricant films with c_f and k_f and discussed their shear rate dependence.

3. Method and materials

3.1. Experimental procedure

We started to oscillate the optical-fiber probe sinusoidally using a piezo actuator. The gap width between the probe tip and the sample was decreased at a constant rate of 2 nm/s using the piezo stage. At an initial state, the probe tip did not make contact with the sample. As the gap was gradually decreased, the probe started to shear the lubricant film on the disk and finally made contact with the substrate. We defined this process as the "approaching process". Subsequently, measurement was conducted in the "separating process", which is a process whereby the gap width was increased at a constant rate of 2 nm/s. In the separating process, the probe first sheared the substrate and then only sheared the lubricant film. Finally, the probe and sample lubricant came off completely. Both in the approaching and separating processes, we measured change in probe amplitude and phase shift, and evaluated c_f and k_f. The oscillation frequency was fixed at 1000 Hz in all of the experiments. The frequency of 1000 Hz was smaller than the resonant frequency of the probe, which was 4945Hz. To discuss shear rate dependence, we conducted each measurement at different oscillating amplitude: 50 nm, 25 nm, 17.5 nm, 10 nm and 5 nm. Corresponding shear rate ranged from 10^2 to 10^6 s^{-1}.

Both in the approaching and the separating processes, gap range whereas the probe tip is shearing lubricant film must be determined. In the approaching process, we defined the gap width at which the probe tip began to contact the lubricant film as the "touch-down gap". Touch-down gap was defined as the point where decrease in the probe amplitude or increase in the phase shift first exceeded noise levels. The noise levels of the measured probe amplitude and phase shift were determined by calculating the threefold variance in the measured data at a time when the probe did not contact the lubricant film. Similarly, we defined the gap width where the probe and the lubricant film completely came off in the separating process as the "take-off gap". Take-off gap was defined as the point where both the increasing probe amplitude and the decreasing phase shift first agreed in the range of noise levels with their average values measured when the probe and the sample were completely apart. These touch-down and take-off gaps represent the gap range whereby the probe end shears the lubricant film and they are not necessarily equal to the film thickness since the lubricant film can be deformed.

3.2. Materials

We used a magnetic disk with a carbon overcoat as a substrate. The surface roughness of the disk was measured by atomic force microscopy. The root-mean-square roughness was 0.45 nm and the peak-to-valley roughness was 1.96 nm. For the sample lubricant we used the non-polar perfluoropolyether lubricant, Fomblin Z03 (Solvey Solexis Inc.). Its chemical formula is,

$CF_3–CF_2(OCF_2CF_2)_n(OCF_2)_mOCF_2–CF_3$.

The molecular weight was 4000 amu and the length of the main chain was around 14 nm. Since the main-chain structure is quite flexible, the Z03 molecule has random coil structures in the bulk state. Dip coating was used to form the molecularly thin lubricant films on the magnetic disk. In dip coating, we prepared the lubricant solution and dipped the disk at a constant speed. When the disk was fully immersed, the disk was pulled out at the same speed as when dipping. Thickness of the film was controlled precisely by the concentration of the solution and the speed of dipping and pulling out. All samples we prepared in this study had a thickness of 2.0 nm, which was verified by an ellipsometer.

4. Experimental result and discussion

Figure 2 shows the experimental results measured with the oscillation amplitude of 25 nm. The longitudinal axes plot the probe amplitude a, phase shift δ and resonant-frequency component a_r. The plotted values of a_r are normalized using their average value measured while the probe and sample did not make contact. The transverse axes plot the gap width between the probe end and the disk. Origins of the gap width were defined as the solid-contact point determined by the increase in the measured resonant-frequency component. The touch-down gap was 4.5 nm and the take-off gap was 150 nm. For the other oscillation amplitudes, quantitatively identical results were obtained (data not shown). Since the thickness of the lubricant film was 2.0 nm, the touch-down gap of 4.5 nm indicates that the contact between the probe tip and lubricant film started at a gap width larger than the thickness of the

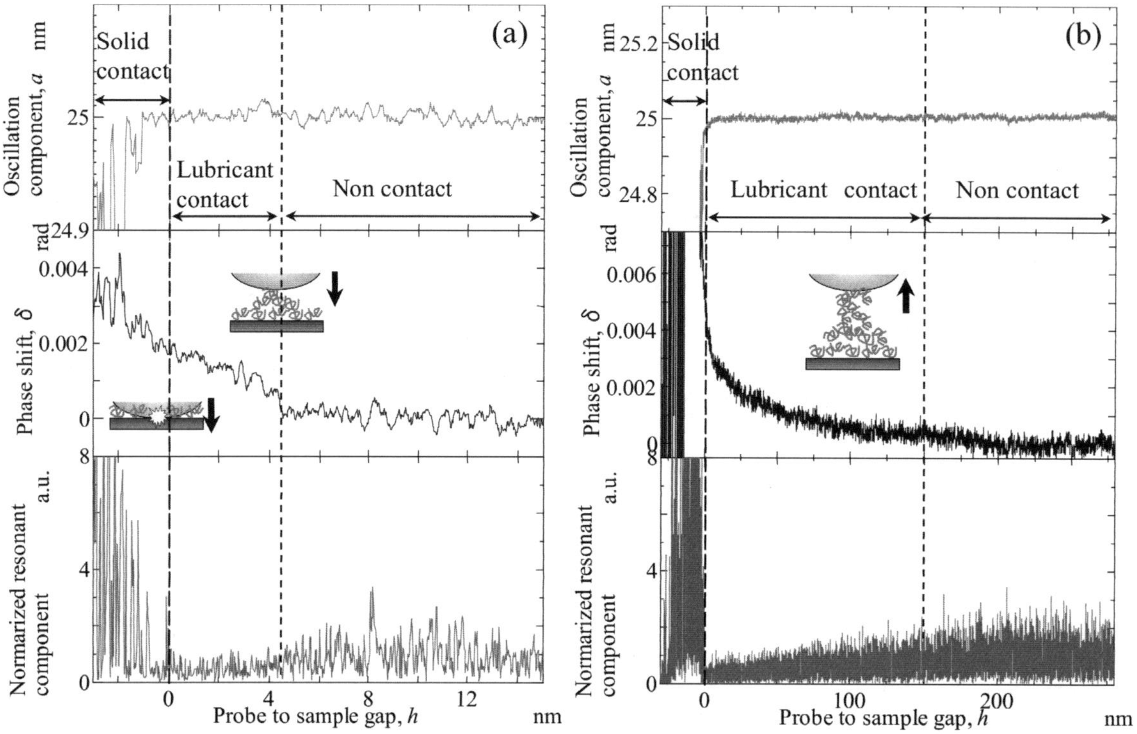

Figure 2. Probe amplitude, phase shift, and normalized resonant component measured with oscillation amplitude of 25 nm in (a) approaching and (b) separating processes.

lubricant film. We considered that the attractive intermolecular force from the approaching probe surface caused the formation of a liquid bridge between the probe tip and the disk. This was not the case with different type of lubricant that strongly adsorbed on to the disk surface [6]. Due to the strong adsorption to the disk surface, the lubricant film cannot be deformed by the attractive force from the probe tip. Therefore, the touch-down gap is equivalent to the film thickness. The take-off gap of 150 nm was much larger than the film thickness. This can be due to the elongation of lubricant gathered near the probe tip during the contact. The formation of liquid bridge in approaching process and its elongation during the separating process are schematically summarized in figure 3. We calculated damping coefficient c_f and elastic coefficient k_f from the measured probe amplitude and phase shift by using equation (1) and (2). Results for different oscillation amplitudes are plotted in figure 4 to 8. The longitudinal axes plot c_f and k_f and the transverse axes plot the gap width. The horizontal dotted lines show the magnitude of noise levels in the determination of c_f and k_f. They were obtained by substituting the threefold variance of the measured probe amplitude and phase shift into equations (1) and (2). The variance was evaluated from the data obtained when the probe did not make contact with the sample. As shown in figure 4 to 8, the elasticity k_f was quite small and comparable to the noise level in most cases. Therefore, we cannot discuss the share rate dependence of k_f in this study. In contrast, the damping coefficient c_f were sufficiently lager than the noise level at the smaller gap widths. We discuss the shear rate dependence of c_f at the gap range where measured values of c_f sufficiently exceeded the noise level, which were the gap widths less than 3 nm in approaching process and the gap widths less than 40 nm in separating process.

The relationships between c_f and shear rates at representative gap width, which are 0.2, 1.0, 2.0, 3.0 nm in the approaching process and 2, 10, 20, 40 nm in separating gap process, (shown as vertical dotted lines in figure 4 to 8) were shown in the figure 9 with double logarithmic plot. Transverse axes show shear rates calculated from the gap widths and the maximum shearing speed, which were

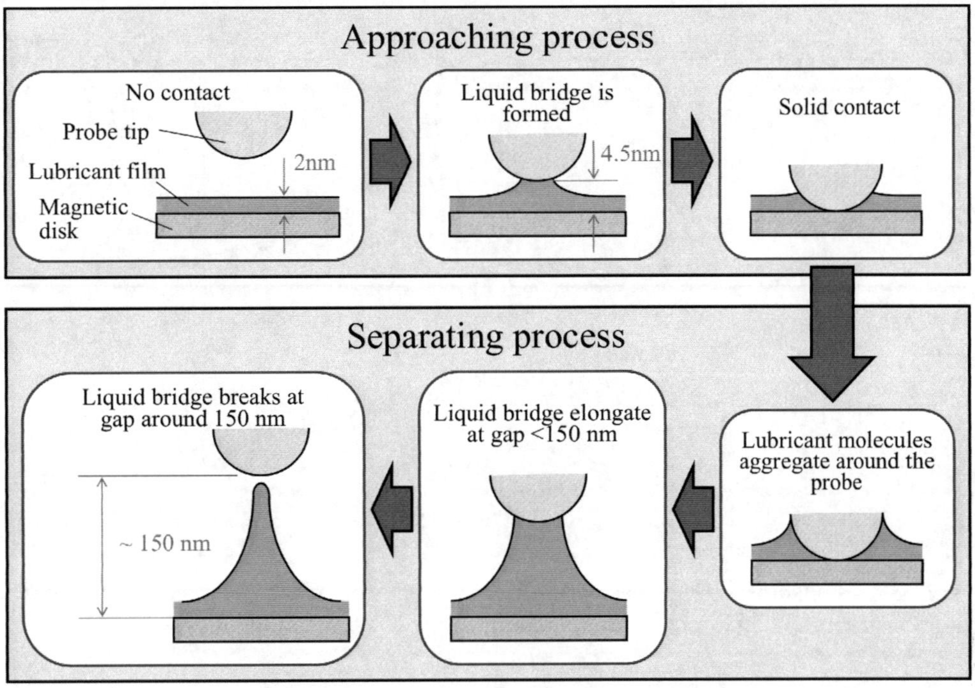

Figure 3. Schematic of configuration of lubricant molecules between the probe tip and the substrate in approaching and separating processes.

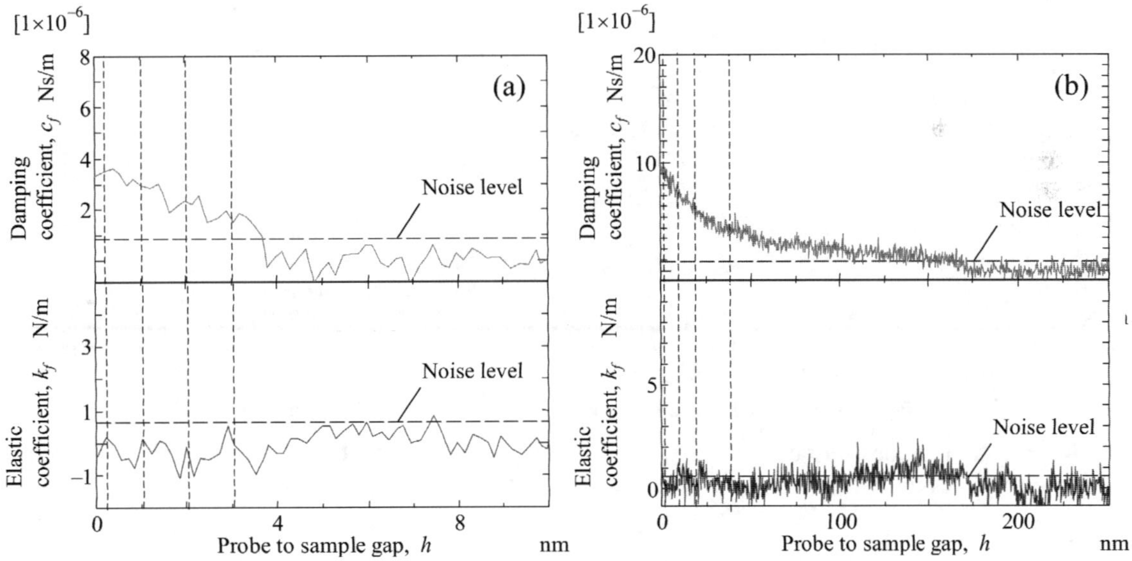

Figure 4. The relationship between gap width and damping coefficient, c_f and elastic coefficient k_f with oscillation amplitude of 50 nm in (a) approaching and (b) separating processes.

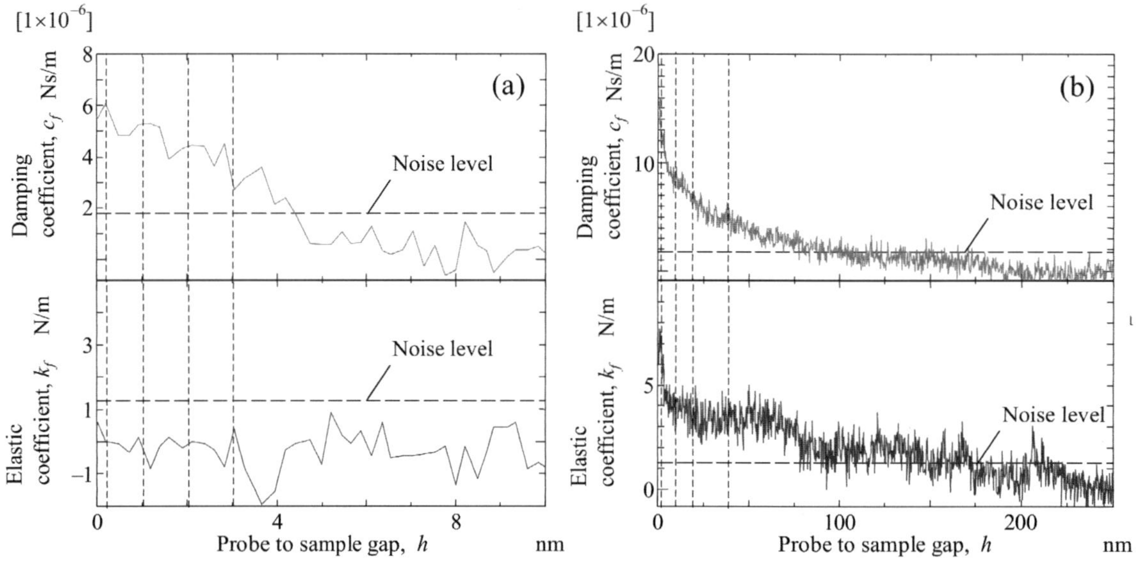

Figure 5. The relationship between gap width and damping coefficient, c_f and elastic coefficient k_f with oscillation amplitude of 25 nm in (a) approaching and (b) separating processes.

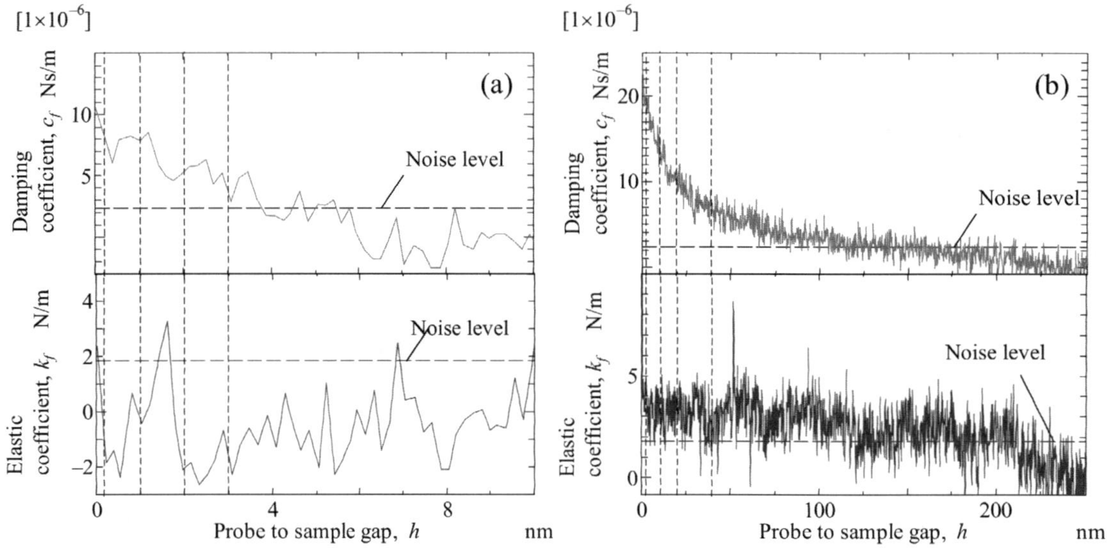

Figure 6. The relationship between gap width and damping coefficient, c_f and elastic coefficient k_f with oscillation amplitude of 17.5 nm in (a) approaching and (b) separating processes.

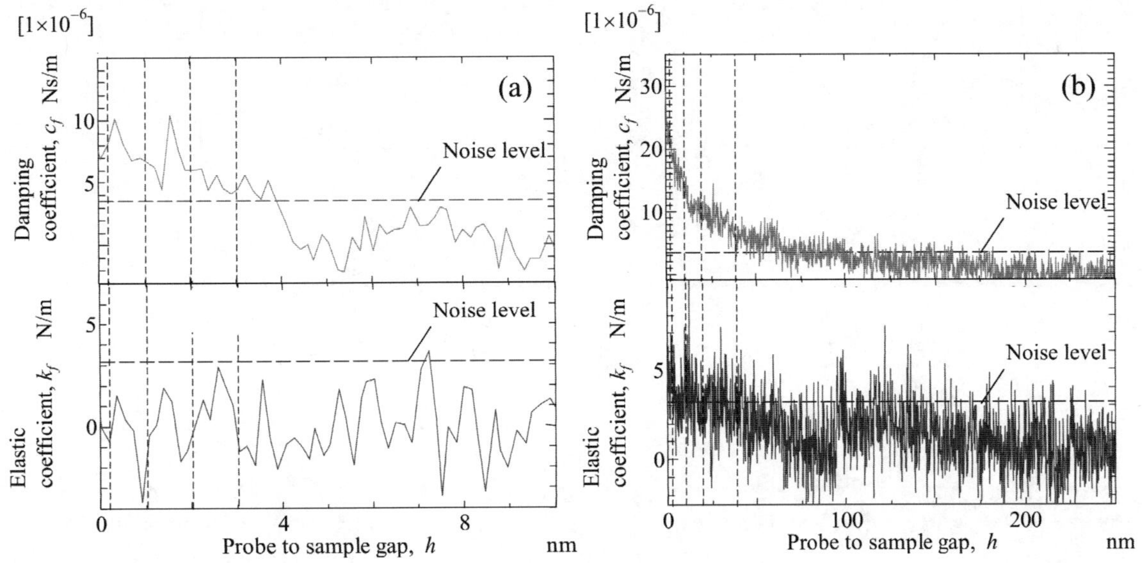

Figure 7. The relationship between gap width and damping coefficient, c_f and elastic coefficient k_f with oscillation amplitude of 10 nm in (a) approaching and (b) separating processes.

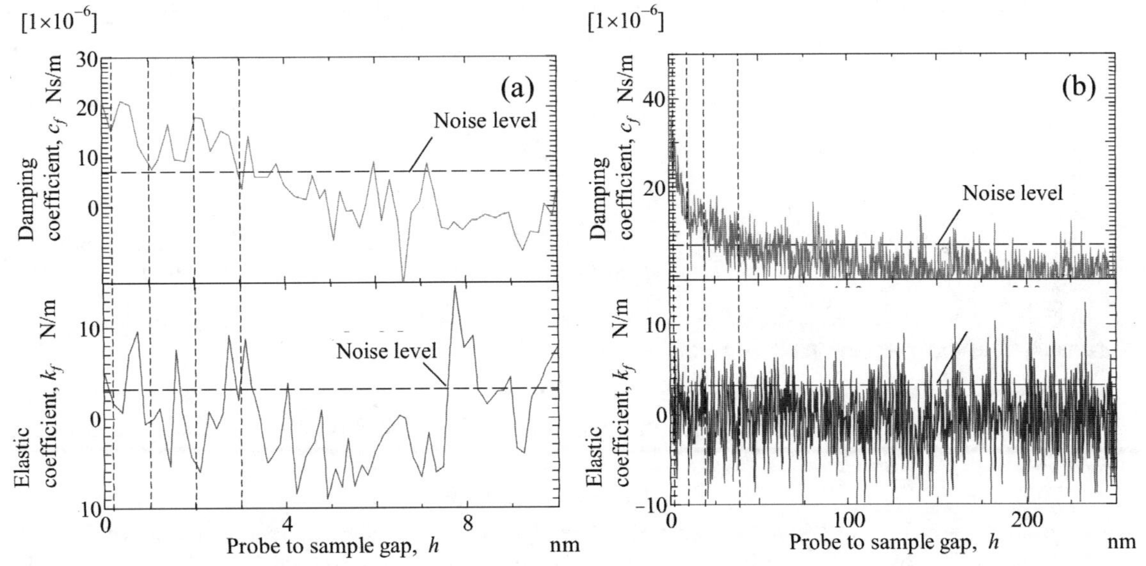

Figure 8. The relationship between gap width and damping coefficient, c_f and elastic coefficient k_f with oscillation amplitude of 5 nm in (a) approaching and (b) separating processes.

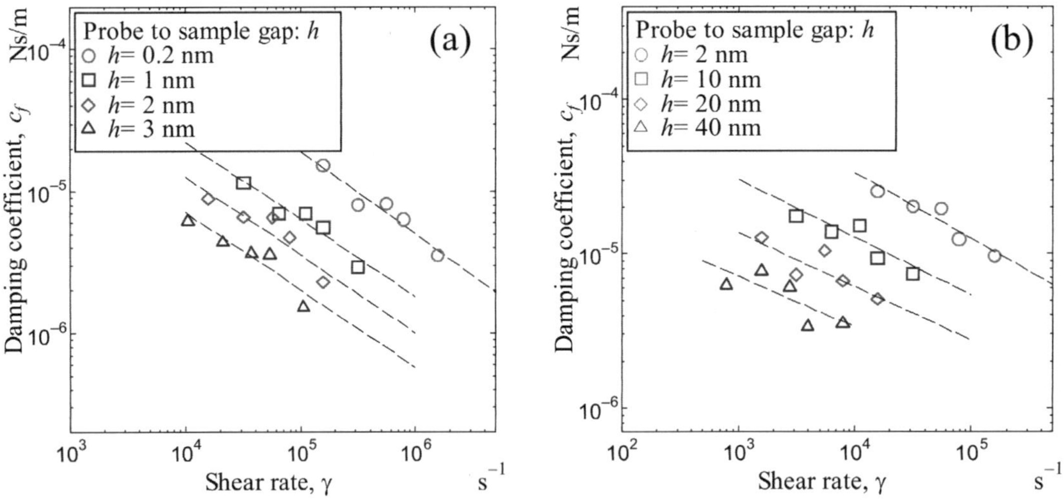

Figure 9. The relationship between shear rate and damping coefficient, c_f, in (a) approaching and (b) separating processes. Both in (a) and (b), damping coefficient decreased exponentially at every gap width. These results indicate the shear thinning behavior of Z03 films on the magnetic disk surface.

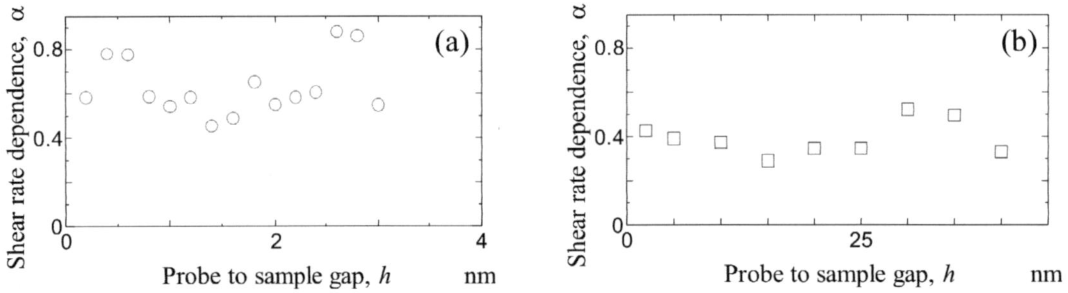

Figure 10. The relationship between shear rate dependence, α, and the gap width, h, in (a) approaching and (b) separating processes.

determined by oscillation frequency and amplitude. Shear thinning behavior is often well expressed by the empirical power law:

$$\log c_f = -\alpha \log \gamma + \beta , \tag{6}$$

where α and β are fitting parameters and γ represents shear rate. The broken lines in figure 9 are the calculation results obtained by fitting the equation (6) to the experimental data. As shown in figure 9, shear rate dependence of c_f was well represented by the empirical power law. These results indicate that shear thinning does occur with molecularly thin Z03 lubricant films coated on the magnetic disk surface. At the other gap width, which should be less than 3 nm in approaching and less than 40 nm in separating, similar shear thinning was observed. We plotted the α against the gap widths as shown in figure 10. The values of α were obtained by the results of a fitting calculation, and they represent the inclination of the fitted line of equation (6). A larger value of α represents a more prominent decrease in viscosity along with increasing shear rate caused by shear-thinning. As shown in figure 10, shear dependence, α ranged from 0.4 to 0.8 in approaching process. On the other hand, it remained constant at around 0.4 in separating process. Although this difference might have some kind of physical origins, future investigations must be performed to explain the details.

5. Summary

To clarify whether or not shear thinning occurs with lubricant films coated on the magnetic disk, we measured the shear rate dependence of the viscoelastic properties of 2 nm-thick lubricant films, using the fiber wobbling method (FWM). In the viscoelastic measurements, oscillation frequency was fixed at 1000 Hz and the oscillation amplitude was varied from 5 nm to 50 nm to change the shear rate. Corresponding shear rates examined in this study ranged from 10^2 to 10^6 s^{-1}. Our experimental results revealed that shear thinning does occur both in the approaching and separating processes at nanometer-sized gap widths. We consider this result is important in the design of head disk interface lubricating systems since the shearing speed dynamically changes during an HDDs' operation.

Acknowledgments

This work was supported in part by SENTAN, the Japan Science Technology Agency (JST), the Storage Research Consortium, the Electro-Mechanic Technology Advancing Foundation, and The Asahi Glass Foundation.

References

[1] Bogy D B and Wu L 2002 *ASME Journal of Tribology* **124** 562-7
[2] Ruth M and Granick S 1997 *Tribology Letters* **7** 161-72
[3] Hahm D C and Bhushan B 1997 *J. Appl. Phys.* **81** 5384-86
[4] Itoh S, Fukuzawa K, Hamamoto Y, Zhang H and Mitsuya Y 2008 *Tribology Letters* **30** 177-89
[5] Itoh S, Fukuzawa K, Hamamoto Y and Zhang H 2010 *Tribology International* (in press)
[6] Hamamoto Y, Itoh S, Fukuzawa K and Zhang H 2010 *Journal of the Japan Society of Mechanical Engineers* **76** 68-79
[7] Yang P C, Chen Y and Vaez-Iravani M 1992 *J. Appl. Phys.* **71** 2499-502
[8] Berg S and Johannsmann D 2003 *Physical Review Letters* **91** 145505
[9] Karrai K and Tiemann I 2000 *Physical Review B* **62** 13174-81
[10] Fukuzawa K, Itoh S, Ando T, Takahashi K, Zhang H and Mitsuya Y 2004 *J. Appl. Phys.* **95** 5189-91
[11] Luengo G, Smoliar L, Grannen K, Marchon B, Jhon M S and Bauer C L 1997 *Macromolecules* **30** 2482-94

Journal of Physics: Conference Series **258** (2010) 012011 doi:10.1088/1742-6596/258/1/012011

Quasi-Static Frictional Test between Silicon Sharp Probes with *in-situ* TEM Observation of Real Contact Point

Tadashi Ishida, Takaaki Sato, Shinsuke Nabeya and Hiroyuki Fujita

Center for International Research on Micro/Nano Mechatronics, Institute of Industrial Science, University of Tokyo, 4-6-1 Komaba, Meguro, Tokyo 153-8505 JAPAN

tadashii@iis.u-tokyo.ac.jp

Abstract. Frictional deformation of a nano-scaled real contact area between silicon MEMS opposing tips was *in-situ* visualized by a transmission electron microscope and the force to deform the real contact area was simultaneously measured. In our experiment, the initial diameter of the real contact point was 6.4 nm. During the frictional motion at 0.01 nm/s, the force applied to the contact point was gradually increased. The maximum frictional force was 61 nN when the diameter of the real contact point was 2.2 nm. After the maximum force, the force gradually decreased to zero because the real contact point was retracted and finally fractured. The friction coefficient was 0.9, which is reasonable in comparison with the previous reports. This result suggested that the friction was caused by the deformation of the real contact point.

1. Introduction

According to Coulomb-Amonton's law, friction is not dependent on apparent contact area at macro scale [1]. This is because the interface between two objects always has small gaps due to many micro/nano asperities on the surfaces. This causes that real contact area is less than the apparent contact area. From this point of view, friction was caused just at the real contact points at micro/nano scale. The real contact area can be assumed by counting the contact points and sum up their contact areas. Some exceptional cases to the law is, however, coming to appear, as nanotechnology developed. In nano- and micro-scaled mechanical devices, which is one of the important application in nanotechnology, have a micro- and nano-scaled contact point [2,3]. At such a small contact point, the interface at the contact point has few or no gaps. This means that the apparent contact area reaches the real contact point. In this condition, friction should depend on the contact area, against the law. Therefore, it is necessary to study nano-scaled tribology.

Additionally, the effect of friction is more significant at the nano scale than at the macro scale [4]. Friction is one of the surface phenomena and propotional to the square of dimension. On the other hand, gravity and inertia are volume phenomena, which are dominant in macro scopic physics. The volume phenomena are proportional to the cubic of dimension. In the miniaturized structure, the surface force becomes more influential than in comparison with the volume force. At the nanoscopic region, friction is one of the most dominant forces to determine the motion of the structure. It is necessary to study nano-scaled tribology for the further development of nano- and micro-scaled mechanical devices [5].

For the study of nano-scaled tribology, it is important to visualize the real contact points, the origins of the friction, during the frictional experiment. To fulfill the requirement, Atomic force microscope (AFM) has been installed in a transmission electron microscope (TEM) [6]. A nano-scaled contact point was visualized during the frictional motion with this equipment [7]. There still remains the problem concerning stability because the AFM apparatus easily suffers from external vibration and drift of piezoelectric actuators. For the further study of nano-scaled tribology, an experimental setup, which allows nano-scopic observation and stable actuation, is necessary. In comparison to the piezoelectric actuators, micro electrostatic actuators have better stability and multi-fuctionalization. Hence, we developed microelectromechanical systems (MEMS)-in- transmission electron microscope (TEM) [8] and performed a frictional experiment under the nanoscopic observation. In this experiment, silicon is chosen for the surface material because it is one of the most popular materials in micro/nano mechanical devices and the devices encountered friction and wear problems [2]. The relationship between frictional force at a single silicon real contact point and its realtime deformation was reported.

2. Experimental Setup

For the visualization of nano-scaled deformation at the silicon real contact point, we performed a frictional experiment using MEMS-in-TEM. Silicon MEMS opposing tips with 2 degree-of-freedom (DOF), MEMS-in-TEM system and a force measurement method between MEMS tips are explained in the following subsections.

2.1. Silicon MEMS Opposing Tips

Figure 1 shows a schematic illustration of a MEMS device with silicon opposing tips. The silicon MEMS opposing tips have electrostatic actuators that allow 2 DOF motion; in the approach direction and the frictional direction. One tip approaches another when a voltage, V_{DX}, is applied between a movable electrode and driving electrodes for approach. Another tip also moves in frictional direction when a voltage, V_{DY}, is applied between a movable electrode and driving electrodes for friction. These electrostatic actuators have precise displacement control within sub-nm order and good stability over several hours.

All the electrodes in this MEMS opposing tips were fabricated with bulk micromachining and opposing tips were formed with focused ion beam (FIB) etching. The detailed fabrication process is explained in ref. 7. Damaged layer was introduced into the surface of opposing tips by FIB etching [9]. To remove this damaged layer, silicon opposing tips were beaten each other in the ultra high vacuum (UHV) chamber of TEM. The damaged layer was removed with this mechanical cleaning. The pure silicon crystalline structures appeared on the surface of tips.

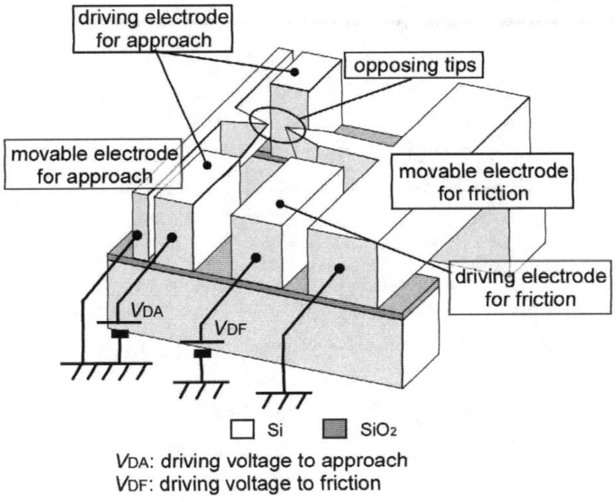

Figure 1. Schematic drawing of MEMS device with silicon opposing tips. Electrostatic actuators between movable electrode and driving electrode move silicon opposing tips to approach and frictional directions.

2.2. MEMS-in-TEM system

Figure 2 shows the MEMS-in-TEM experimental setup. The MEMS device with silicon opposing tips were inserted into UHV-TEM specimen chamber using a specially customized sample holder for MEMS device. This TEM sample holder had 9 feedthroughs to apply voltage to the electrostatic actuators. With changing the voltage, the motions of opposing tips were controlled, resulting in the frictional test inside TEM. The spatial resolution was 0.2 nm and the time resolution was 1/30 s. The vacuum of TEM specimen chamber was $3x10^{-8}$~$1x10^{-7}$ Pa, which kept the surface of tips clean. An electron beam from a TEM electron gun irradiated silicon opposing tips and projected an high resolution image on its screen. Strong magnetic field over 1 T was applied to the device due to electron lens of TEM. The current density of the TEM electron beam for the observation was $1.6x10^4$ A/m^2. With this electron beam, the heating effect was less than 1 K, which was enough low to consider that the condition was at room temperature [10].

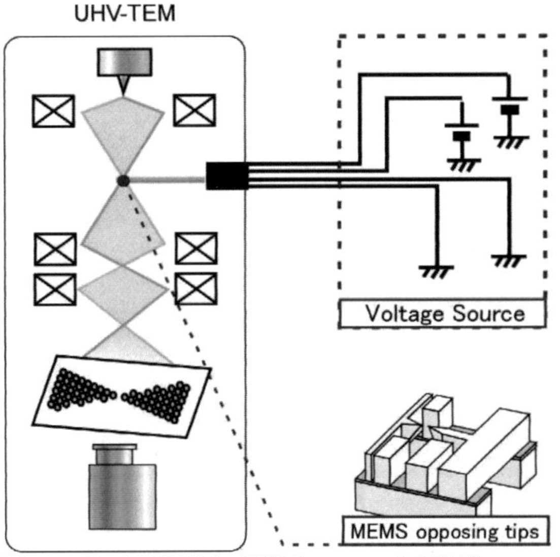

Figure 2. MEMS-in-TEM system. MEMS opposing tips can be driven inside TEM specimen chamber, applying driving voltages with a TEM special sample folder for MEMS device.

2.3. Force Measurement Method between MEMS Tips

The frictional force can be calculated using displacement difference of the movable tip between with and without a contact at a certain driving voltage. The displacement before the contact is determined by the balance between an electrostatic force and a restoring force of the movable electrode. After the contact, the displacement is determined by the balance among an electrostatic force, a restoring force and a frictional force. The frictional force is always applied to the contact point against the moving direction. Therefore, the displacement becomes smaller by this frictional force. According to the Hook's law, the frictional force between tips can be calculated by the product between the displacement difference and the spring constant of the movable electrode. The displacement of the movable electrode can be measured by TEM observation. The spring constant of the movable electrode can be calculated from its design [11]. Using the same method, a nominal force applied to the contact point can be obtained. The spring constants of the movable electrodes for approach and for friction are 60 N/m, 23 N/m, respectively. Considering that the spatial resolution of our TEM is 0.2 nm, the force resolution should be 12 nN in approach direction and 4.6 nN in frictional direction.

3. Frictional Experiment between Silicon Tips

Figure 3 shows the experimental procedure of the friction test between silicon tips. The movable tip for approach moves towards the movable tip for friction (Fig.4a). The tips are brought into contact. After the contact, a silicon nanocontact is formed between tips [12]. The movable tip for friction is driven, resulting in the deformation of the silicon nanocontact with frictional motion (Fig. 4c). The silicon nanocontact is fractured with the frictional motion of the tip (Fig. 4d). These steps are performed in the TEM specimen chamber to observe its deformation during the experiment.

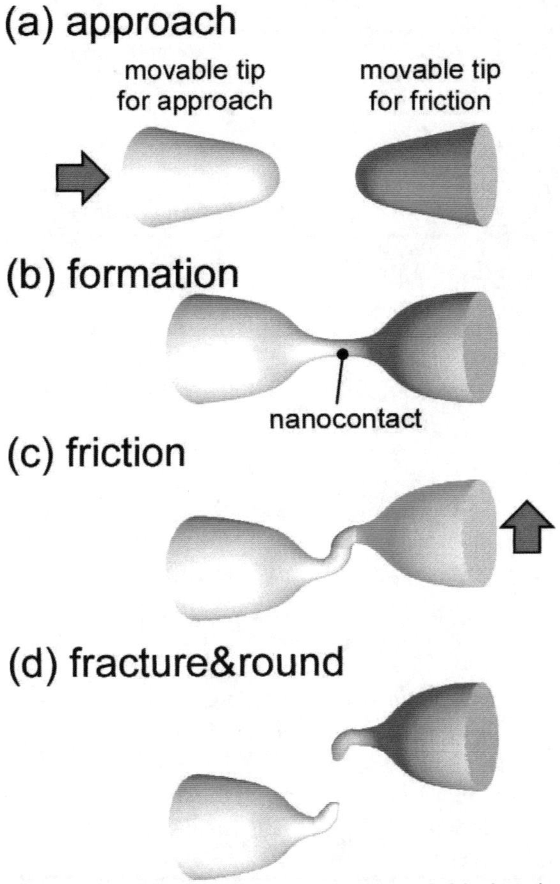

Figure 3. Experimental procedure of frictional test between silicon opposing tips. (a) approach, (b) formation of silicon nanocontact, (c) frictional deformation of the silicon nanocontact, (d) fracture of the silicon nanocnatact and rounding of the corner.

4. Deformation of Silicon Nanocontact and its Frictional Property

Figure 4 shows the nanometer scaled deformation of the silicon nanocontact during the experiment. The silicon nanocontact of 1.3 nm in diameter was formed between tips at the moment that the movable tip for approach pushed the tip for friction by 1.1 nm. The silicon nanocontact grew in diameter by a silicon atomic surface diffusion, while the actuator position was kept stationary. After 30 min from the contact, the diameter of the silicon nanocontact was 6.4 nm (Fig. 4a).

The silicon nanocontact formed between tips was plastically deformed with frictional motion of the tip. The displacement speed of the tip for friction was 0.01 nm/s in average during the frictional motion. The diameter of the silicon nanocontact gradually decreased from 6.4 nm to 3.6 nm before the displacement was 5.8 nm. At 8.7 nm in displacement, the diameter of the silicon nanocontact was 2.4 nm (Fig. 4b). The silicon nanocontact was finally fractured at the displacement of 16.2 nm (Fig. 4c).

The corners were rounded from 0.6 nm to 1.1 nm in radius of curvature in 5 min (Fig. 4d). During this frictional experiment, the displacement with the silicon nanocontact was smaller in comparison to that without the silicon nanocontact.

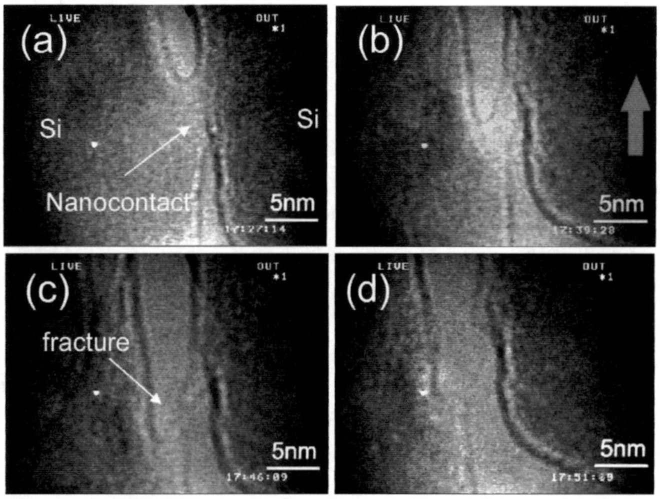

Figure 4. TEM images of each experimental step. (a) formation, (b) frictional deformation, (c) fracture, (d) rounding of the silicon nanocontact.

5. Discussion

From the observation of the displacement difference of the tip, the force applied to the silicon nanocontact was calculated. The force did not increase so much from the initial position to 5.8 nm in displacement. After the displacement exceeded 5.8 nm, the force suddenly increased. The force reached the maximum value of 61 nN with 8.7 nm in displacement. After the maximum value, the force gradually decreased to zero. At the moment when force became zero, the silicon nanocontact was fractured.

In this experiment, the normal force at the contact point was 66 nN. On the other hand, the maximum force applied to the silicon nanocontact during the frictional motion was 61 nN. From these values, the frictional coefficient was calculated to 0.9 using the equation, $\mu = F/N$, where μ is friction coefficient, F is friction force and N is normal force. According to the previous experimental reports, the static friction coefficient this experiment was comparable value [13-16]. This result suggested that the deformation of the silicon nanocontact caused the friction between tips.

6. Conclusions

The silicon nanocontact was plastically deformed caused by the frictional motion using silicon MEMS opposing tips under TEM observation. From the observation, it was confirmed that the frictional force was caused by the deformation of the real contact point during the frictional motion. This MEMS-in-TEM system will be a powerful tool to study nano-scaled tribology.

7. Acknowledgement

This work was supported by KAKENHI; the grant-in-aid of Specially Promoted Research 21000008 sponsored by Japan Society for the Promotion of Science (JSPS).

References

[1] Bowden F P and Tabor D 1967 *Friction and lubrication*, (Mehuen & Co.)
[2] Gabriel K J, Behi F and Mahadevan R 1990 *Sens. Actutors A Phys.* **21-23** 184
[3] Jensen B D, Chow L L-W, Kuangwei H, Saitou K, Volakis J L, Kurabayashi K 2005 *J. Microelectromech. Sys.* **14** 935

[4] Y. –P. Zhao, L. S. Wang and T. X.Yu, J. 2003 *Adhesion Sci. Technol.* **17** 519
[5] Tambe N S and Bhushan B 2004 *Nanotech.* **15** 1561
[6] Kizuka T, Ohmi H, Sumi T, Kumazawa K, Deguchi S, Naruse M, Fujisawa S, Sasaki S, Yabe A and Enomoto Y 2001 *Jpn. J. Appl. Phys.* **40** L170
[7] Fujisawa S and Kizuka T 2003 *Tribol. Lett.* **15** 163
[8] Ishida T, Nakajima Y, Kakushima K, Mita M, Toshiyoshi H and Fujita H 2010 *J. Micromech. Microeng.* **20** 075011
[9] Kato N I, Kohno Y and Saka H 1999 *J. Vac. Sci. Technol. A* **17** 1201
[10] Iijima S 1987 *Microclusters* ed. Sugano S, Nishina Y and Ohnishi S (Springer-Verlag) pp 186-192.
[11] Rebeiz G M 2002 *RF MEMS Theory, Design, and Technology* (New York: Wiley) 21–30.
[12] Ishida T, Kakushima K, Sasaki N and Fujita H 2010 *Nanotech.* **21** 435705 (5pp)
[13] Deng K and Ko W H 1992 *J. Micromech. Microeng.* **2** 14
[14] Bhushan B and Li X 1997 *J.Mater.Res.* **12** 59
[15] Schirmeisen A, Jansen L, Hölscher H and Fuchs H 2006 *Appl. Phys. Lett.* **88** 123108
[16] Alsem D H, Dugger M T, Stach E A, and Ritchie R O 2008 *J. Microelectromech. Sys.* **17** 1144

Journal of Physics: Conference Series **258** (2010) 012012 doi:10.1088/1742-6596/258/1/012012

Formation of Iron Carbide Nanorod by Pulsed Plasma Chemical Vapor Deposition

S Yooyen, T Kawamura, S Kotake and Y Suzuki
Graduate School of Engineering, Mie University
1577 Kurimamachiya-cho, Tsu City, Mie Prefecture, Japan 514-8507

E-mail: soemsak@qm.mach.mie-u.ac.jp

Abstract. Iron carbide nanorods were produced on the surface of pure iron sheet by means of pulsed plasma chemical vapor deposition. Hydrogen plasma was used in the initial state, varied from 10 to 60 minute, for the purpose of cleaning and preparing nanoparticles. The formation was done under methane atmosphere, varied from 2 to 30 minute. X-ray diffraction (XRD) patterns analysis showed that iron carbide was found on all of these experiment conditions. By field emission scanning electron micro spectroscopy (FESEM) observation, iron carbide nanorods, however, were found neatly only on the methane plasma treating at 10 minute.

1. Introduction

Iron-based materials such as ferrite, austenite, martensite and iron carbides Fe_3C have been extensively studied because of their various industrial applicable. As in bulk materials, Fe_3C nanopartilces suitable for magnetic applications were fabricated by chemical vapor condensation (CVC) [1] which spent 2 hour in fabrication. The fabrication time can be reduced by laser deposition which Fe_3C films were produced [2]. Among the long fabrication time of CVC and medium vacuum level of laser deposition, plasma enhanced chemical vapor deposition (PECVD) which is one of cost-effective methods and controllability of process parameters as well as ease to scale up is proposed. Carbon nanotubes were grown by this method under the deposition period less than 1 hour [3-4]. Diamond-like carbon (DLC) films were deposited by pulsed discharge plasma CVD in the rough vacuum level [5]. To synthesis nanotubes or nanorods, catalyst layers were deposited on substrate to support the growth of nanoparticles [6]. The growth mechanism from the effect of catalytic metal decomposes the acetylene gas into carbon and hydrogen by plasma enhancement was summarized by Wunderlich [7]. In this study, the synthesis was done directly on pure iron substrate which was acted as both the catalyst and supporter for the growth of nanoparticles. The synthesis was proposed to be done under rough vacuum level with deposition time under 1 hr. The surface morphology was observed by FESEM while the crystal structure was evaluated by XRD.

2. Experimental details

The synthesis was done by means of pulsed discharge plasma chemical vapor deposition (Figure 1.). Discharge frequency was 1000 Hz, with discharge duty 20%, discharge voltage 600 volt and discharge distance 10 mm. The substrate was polished with SiC paper and cleaned by ultrasonic degreasing before placed in the vacuum chamber. The substrate temperature was maintained about

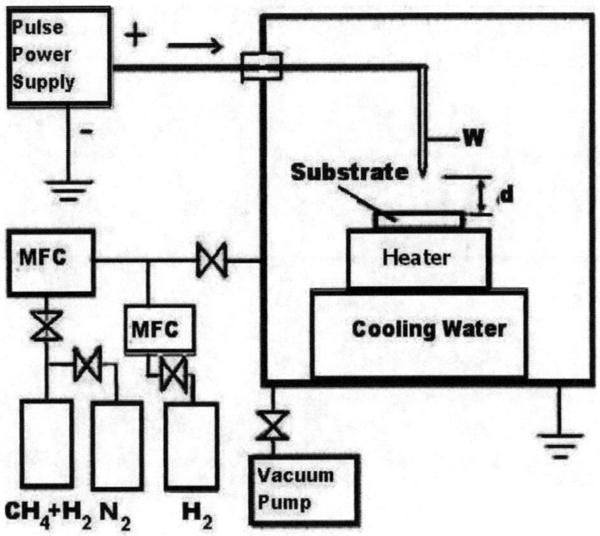

Figure 1. Schematic diagram of experimental tool

600 °C by a resistive heater in a vacuum chamber. The chamber was evacuated to the initial pressure. To prepare nanoparticles on pure Fe substrate, the preheat process was done under H_2 plasma. Nanorods or nanotube can be grown from these nanoparticles [5-6]. The synthesis using CH_4:H_2 as a carbon source was done under a pressure of 4 kPa [8]. The preheat process was varied from 10 to 60 minutes while CH_4:H_2 was supplied varied from 2 to 30 minute [Table 1.]. The surface morphology was observed by FESEM. XRD technique was used to evaluation.

Table 1. Experiment time (minute)

H_2	CH_4:H_2
10	10
10	30
10	60
10	2
30	10
30	30
60	10

3. Result and discussion

From FESEM evaluation, when the substrate was treated by hydrogen plasma at 10 minute [Figure 2.], nanoparticles were found on the substrate [6]. These substrates were supplied methane plasma varied at 10, 30 and 60 minute. Nanorods were found on substrates which supplied methane plasma at 10 minute [Figure 3.]. Further supplied methane plasma, the rods became fiber and the amount of rods trend to be reduced [Fig.4]. It may be because of bombardment effect [9]. After the rods grew by enhancement of electrons emitted from substrate surface and bombardment, bombardment itself later destroyed the rods when further the proper time [7]. The same result was found when hydrogen plasma was treated at 30 minute which nanorods were rarely found on the 60 minute of methane plasma supplied. The proper time to synthesis nanorod was 10 minute while hydrogen preheated for 10 minute was enough for the growth of rods. The size of the rod was about 60 nanometer in diameter and about 600 nanometer in length [Figure 5.].

Figure 2. Preheat by H_2 10 minute

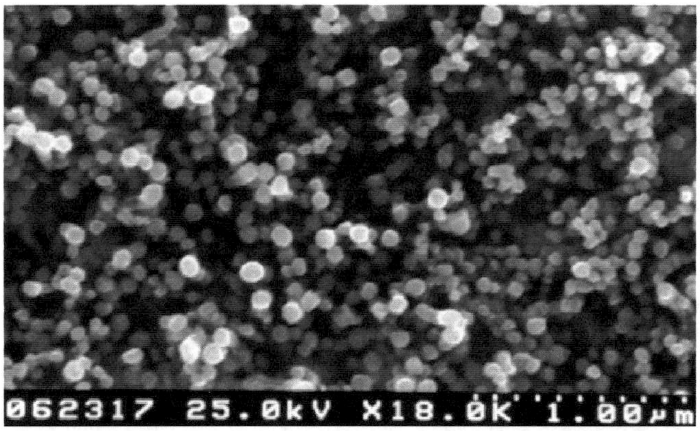

Figure 3. H_2 10 minute and CH_4:H_2 10 minute

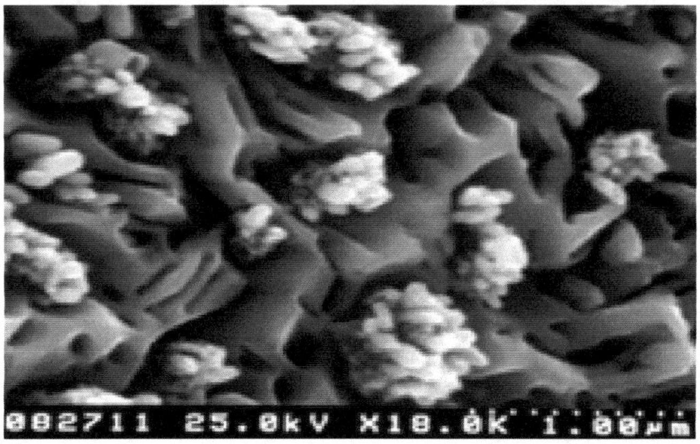

Figure 4. H_2 10 minute and CH_4:H_2 60 minute

Figure 5. Nanorod, d ≈ 60 nanometer and l ≈ 600 nanometer

The samples found nanorods were evaluated by XRD. The diffraction of the difference hydrogen preheated treatment at the same methane treatment period showed the same peak. The main 3 intense peaks were the diffraction of Fe at 44.36 (1,1,0), 64.55 (2,0,0) and 81.92 (2,1,1). The other peaks were well match with iron carbine diffraction [10]. The synthesized rods were supposed to be iron carbine nanorod [Figure 6.].

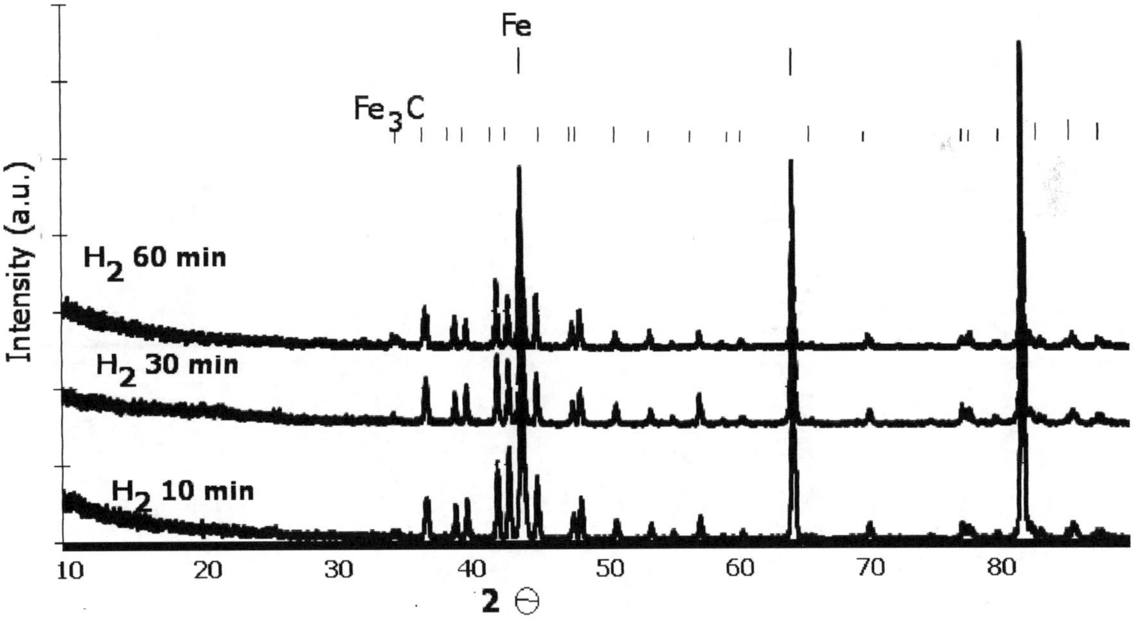

Figure 6. XRD of hydrogen preheated treatment at the same methane treatment.

4. Conclusion

Iron carbide nanorods have been synthesized by means of pulsed plasma chemical vapor deposition. The optimum time for hydrogen plasma preheated was about 10 minute as same as a time for methane plasma. The size of the rod estimated from FESEM in higher magnification was about 60 nanometer in diameter and about 600 nanometer in length. XRD patterns analysis showed that iron carbide was found on all of this experiment conditions.

5. Acknowledgements

The authers would like to The authors would like to thank the Instrumental Analysis Center, Mie University for FESEM and XRD support.

6. References

[1] D W Lee, J H Yu, B K Kim, and T S Jang. *J.Alloys and compounds* **449** (2008) 60-64
[2] Y Yamada, H Yoshida, K Kouno, and Y Kobayashi. *JPCS*, **217** (2010) 012096
[3] M Jonsson, O A Nerushev, and E E B Cambpbell. *Appl. Phys.* A **88**, 261-267 (2007)
[4] H Sato, T Sakai, A Suzuki, K Kajiwara, K Hata, and Y Saito. *Vacuum* **83** (2009) 515
[5] M Noda and M Umeno. *Diamond & Related Materials* **14** (2005) 1791
[6] H Sato and K Hata. *NDFCT.* **16** No. 3 (2006) 163
[7] W Wunderlich. *Diamond & Related Materials* **16** (2007) 369
[8] S Yooyen, S Kotake, Y Suzuki and M Noda. *Proc. 25th ITC-CSCC* 2010, 903.
[9] J García-Céspedes, M Rubio-Roy, M C Polo, E Pascual, J L Andújar and E Bertran. *Diamond & Related Materials* **16** (2007) 1131
[10] D Chaira, B K Mishra and S Sangal. *Powder Technology* **191** (2009) 149-154

International Conference on Science of Friction 2010 (ICSF2010) IOP Publishing
Journal of Physics: Conference Series **258** (2010) 012013 doi:10.1088/1742-6596/258/1/012013

Unique Near-Zero Friction Regime of C_{60} Molecular Bearings Along [12$\bar{3}$0] Direction

Noriaki Itamura[1], Hiroko Asawa[1], Kouji Miura[2] and Naruo Sasaki[1]

[1] Department of Materials and Life Science, Faculty of Science and Technology, Seikei University, 3-3-1 Kichijoji-Kitamachi, Musashino-shi, Tokyo 180-8633, Japan

[2] Department of Physics, Aichi University of Education, Hirosawa 1, Igayacho, Kariya-shi, Aichi 448-8542, Japan

E-mail: naru@st.seikei.ac.jp

Abstract. We numerically analyzed unique near-zero friction regime of the C_{60} molecular bearings, graphite/C_{60}/graphite interface, for the lateral scan along the [12$\bar{3}$0] direction under the relatively low loading condition. Here the C_{60} molecule slides, facing its six-membered ring nearly parallel to both the upper and lower graphene sheets. The sinusoidal motion of the C_{60} molecule along the carbon bond is continuous and reversible during the forward and backward scans. As a result, the hysteresis loop of the lateral force curve nearly disappears, which leads to a mean frictional force of nearly zero, $\langle F_L \rangle \simeq 0$. The mechanism of this conservative motion is clarified by comparing the structural optimization of the C_{60} molecular bearing system with the direct calculation of the local minimum position located on the total potential energy surface V_{total}. The energy barrier between the neighboring minimum positions always exists, which prevents the C_{60} molecule from taking stick-slip motion.

1. Introduction

Control of the superlubricity leads to the energetically effective control of the movement of objects at the interface at from micro- to macro-scopic scale. As a first step to control superlubricity, we have paid attentions to the combination of the fullerene and graphene, and developed several types of the fullerene molecular bearing system, where the fullerene/graphene interface plays a dominant role in superlubricity [1, 2, 3, 4, 5]. Next we numerically studied the anisotropy of the superlubricity of the C_{60} bearing system by molecular mechanics simulation [6]. It is clarified that, under the low average loading force of $\langle F_z \rangle$ =0.27 nN corresponding to the graphene interlayer distance d=1.3 nm, the simulated mean lateral force $\langle F_L \rangle$ takes a maximum peak value along the [10$\bar{1}$0] scan direction, and a minimum value of nearly zero less than 1 pN along the [12$\bar{3}$0] scan direction. The peak of $\langle F_L \rangle$ along the [10$\bar{1}$0] scan direction is derived from the stick-slip motion of the C_{60} molecule due to the commensurate stacking of the six-membered rings between the C_{60} molecule and the graphene sheet [7, 8, 9]. Then we clarified that, at the peak region, the tilting motion of the C_{60} molecule and the elastic contact at the C_{60}/graphene interface contribute to the superlubricity of the C_{60} molecular bearings [9]. However, the mechanism of the minimum value of $\langle F_L \rangle$ along the [12$\bar{3}$0] scan direction is not discussed in detail yet. Therefore, in this work, we studied the unique near-zero friction regime along the [12$\bar{3}$0] direction, and analyzed the mechanism of the conservative continuous motion of the C_{60} molecule by calculating the total potential energy surface.

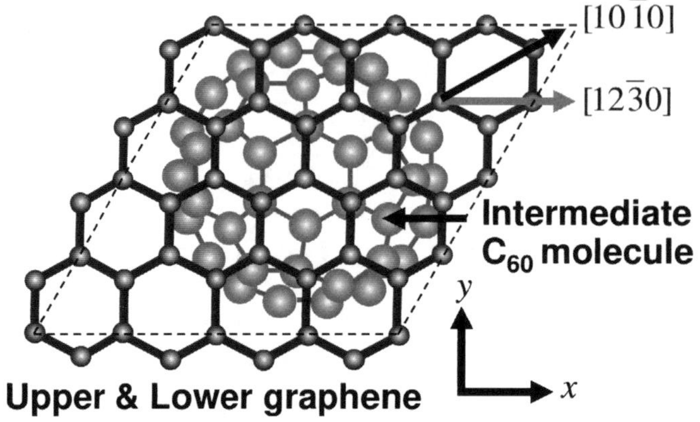

Figure 1. Model of the C_{60} bearing system (graphene/C_{60}/graphene interface). The blue upper graphene sheet and the red intermediate C_{60} molecule are illustrated with the broken parallelogram of the 1×1 unit cell. The x direction corresponds to the $[12\bar{3}0]$ scanning direction.

2. Model and Method of Simulation

In the simulation, we adopted the same model and method of simulation as those used in our previous works [4, 5, 6, 7, 8, 9]. Fig. 1 shows the model of the C_{60} bearing system with the closely packed structure of the C_{60} molecules [1, 2, 3, 4, 5, 10]. Only the intercalated C_{60} molecule sandwiched by the upper and lower rigid graphene sheets is assumed to relax. The periodic boundary condition is applied not to the z-direction but to the 1×1 unit cell of the C_{60} bearing system within the (0001) plane [Fig. 1]. The Tersoff potential $V_{\rm cov}$ [11] and the modified Lennard-Jones potential $V_{\rm vdW}$ [12, 13] are used as the covalent bonding and nonbonding interaction energies, respectively. Then, the lateral position L of the upper rigid graphene sheet is varied along the $[12\bar{3}0]$ and its backward directions, with a fixed interlayer distance $d = 1.3$ nm between the upper and lower rigid graphene sheets, which results in the mean vertical loading force of $\langle F_z\rangle = 0.27$ nN. Here the initial structure is set so that the AB-stacking registry of the six-membered ring between the graphene sheets and the intermediate C_{60} molecule is satisfied. For each scanning position L, the metastable structure of the C_{60} molecular bearing system is obtained by minimizing the total energy $V_{\rm total} = V_{\rm cov} + V_{\rm vdW}$ using the Polak-Rebiere-type conjugate gradient (CG) method [14] under the convergence criterion that the maximum absolute force acting on all the relaxed atoms should become lower than 1.6×10^{-2} pN. Thus the lateral force opposite to the scan direction F_L acting on the upper graphene sheet for each 1×1 unit cell is obtained for $0 \le L \le 0.506$ nm during the forward and backward scans.

3. Results

In this section, the lateral force curve and the C_{60} dynamics are shown, which are explained by analyzing the total potential energy $V_{\rm total}$ for each scan position L of the upper graphene sheet.

3.1. Lateral force curve

Fig. 2 shows the lateral force F_L curve plotted as a function of the scan position L for the forward ($[12\bar{3}0]$ direction) and backward scans. F_L clearly exhibits a sinusoidal behavior with a lattice period of $a = 0.253$ nm along the $[12\bar{3}0]$ scan direction, and F_L curves for both the forward and backward scans are completely the same as each other. Therefore the hysteresis loop does not appear in the force curve, where the average lateral force $\langle F_L\rangle$ exhibits nearly zero, $\langle F_L\rangle \simeq 0$.

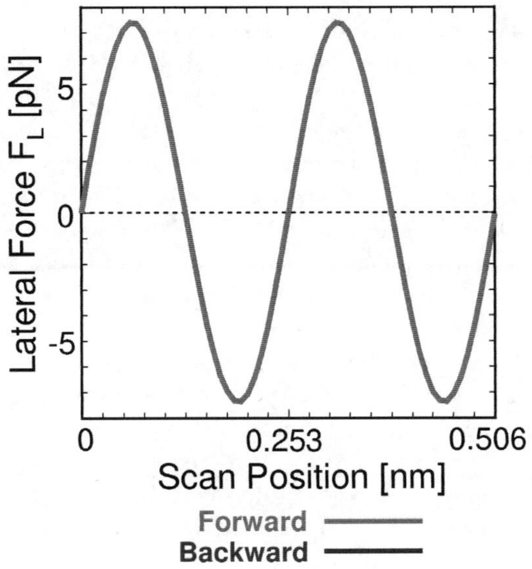

Figure 2. Lateral force curve of the C_{60} bearing system along $[12\bar{3}0]$ $(= x)$ direction for 0 nm $\leq L \leq 0.506$ nm for $\langle F_z \rangle = 0.27$ nN. The forward (red) and backward (blue) scans are completely the same as each other.

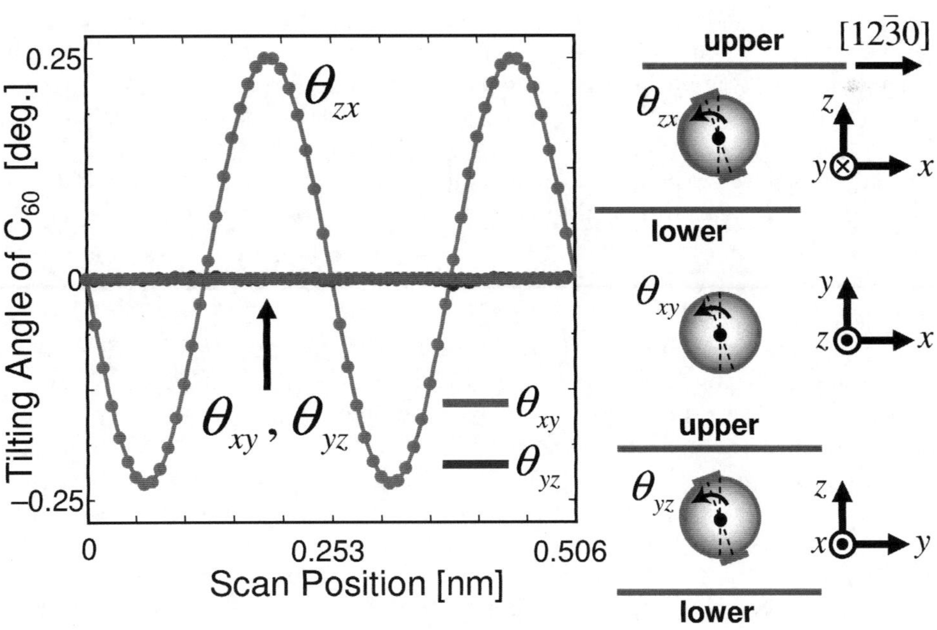

Figure 3. Tilting angles of the C_{60} molecule within the $z-x$, $x-y$, and $y-z$ planes, θ_{zx}(red), θ_{xy}(green), and θ_{yz}(blue), respectively, plotted as a function of the scan position L of the upper graphene sheet along the $[12\bar{3}0](= x)$ direction for $\langle F_z \rangle = 0.27$ nN.

3.2. Dynamics C_{60} molecule

The mechanism of the nearly zero friction can be ascribed to the motion of the C_{60} molecule during the scanning process. Since the interlayer distance $d=1.30$ nm considered in the present work is near an equilibrium distance $d = 1.32$ nm of the graphite/C_{60}/graphite interface as discussed in refs. 7 and 8, C_{60} molecule deforms quite little even during the scanning process. Therefore the tilting angles of the C_{60} molecule within the $x-y$, $y-z$ and $z-x$ planes, θ_{xy}, θ_{yz} and θ_{zx}, are very small, respectively, as shown in Fig. 3. The maximum tilting angles are at most $|\theta_{xy}| \simeq 0.002°$, $|\theta_{yz}| \simeq 0.007°$ and $|\theta_{zx}| \simeq 0.25°$. This means that the C_{60} molecule slides facing its six-membered ring nearly parallel to the upper and lower graphene sheets during the scanning process. Therefore motion of the upper six-membered ring is equivalent to that of the C_{60} molecule itself.

As shown in Figs. 4(a)-4(g), the six-membered ring slides along the path $C_1 \rightarrow C_2 \rightarrow C_3$, when the upper graphene sheet is scanned forward along the $[12\bar{3}0]$ $(= x)$ direction within $0 \leq L \leq 0.506$ nm as follows: First the upper and lower graphene sheets take the AA-stacking registry for $L=0$ nm and the six-membered ring of the C_{60} molecule is initially located above the AB stacking site, C_1 (Fig. 4(a)). As the upper graphene sheet is scanned forward along the $[12\bar{3}0]$ direction, the six-membered ring of the C_{60} molecule slides nearly along the carbon bond C_1-C_2 of the lower graphene sheet (Figs. 4(b)-4(d)). When the upper graphene sheet slides for $L = 0.253$ nm, one period of lattice along the $[12\bar{3}0]$ direction, the upper and lower graphene sheets take the AA-stacking registry again, and the six-membered ring of the C_{60} molecule becomes located above the AB stacking site, C_2 (Fig. 4(e)). Considering the symmetry, the structure of Fig. 4(e) is essentially equivalent to that of Fig. 4(a). Then, for $0.253 \leq L \leq 0.506$ nm, the six-membered ring of the C_{60} molecule takes a sliding motion nearly along the carbon bond C_2-C_3 (Figs. 4(e)-4(g)) similar to that nearly along C_1-C_2 for $0 \leq L \leq 0.253$ nm (Figs. 4(a)-4(e)). Finally, for $L = 0.506$ nm, two periods of lattice along the $[12\bar{3}0]$ direction, the upper and lower graphene sheets take the AA-stacking registry, and the six-membered ring of the C_{60} molecule becomes located above the AB stacking site, C_3 (Fig. 4(g)). The structure of Fig. 4(g) is completely equivalent to that of Fig. 4(a). For the backward scan along $-x$ direction, the six-membered ring of the C_{60} molecule takes a sliding motion $C_3 \rightarrow C_2 \rightarrow C_1$ just opposite to that for the forward scan. As shown in Fig. 4(h), trajectories for the forward and backward scans are completely the same as each other. This means the C_{60} motion is reversible and continuous for the forward and backward scans of the upper graphene sheet along $[12\bar{3}0]$ $(=x)$ direction. Thus the C_{60} molecule slides nearly along the carbon bonds and exhibits sinusoidal zigzag motion among C_1, C_2 and C_3.

3.3. Total potential energy surface

Figs. 5(a)-5(g) show the total potential energy surface $V_{\text{total}}(x, y; L) = V_{\text{cov}} + V_{\text{vdW}}$ plotted as a function of the center position (x, y) of the C_{60} molecule within x-y plane, for each upper graphene position L. The simulation procedure is as follows: For each scan position L of the upper graphene sheet, the C_{60} molecule is scanned within the square region (0 nm $\leq x \leq 0.6$ nm, $0 \leq y \leq 0.6$ nm) to calculate V_{total}. Here it is noted that all the optimized coordinates of the 60 carbon atoms of the C_{60} molecule for each scan position L are used to calculate V_{total}. As a result the C_{60} molecule (or the center position of the C_{60} molecule) is located at the local minimum point P indicated by the larger circle as illustrated in Fig. 5.

As shown in Fig. 5(a), $V_{\text{total}}(x, y; L)$ initially exhibits the two-dimensional pattern reflecting the honeycomb network of the graphene sheet. The C_{60} position P corresponds to the AB-stacking site C_1 between the C_{60} molecule and the upper graphene sheet, located on the vertices of the hexagon. The blue regions around the carbon bond correspond to the low-energy regions. On the other hand, the local maximum points are the A-A stacking sites. These maximum points are located on the center positions of the hexagon (hollow sites) and are surrounded by

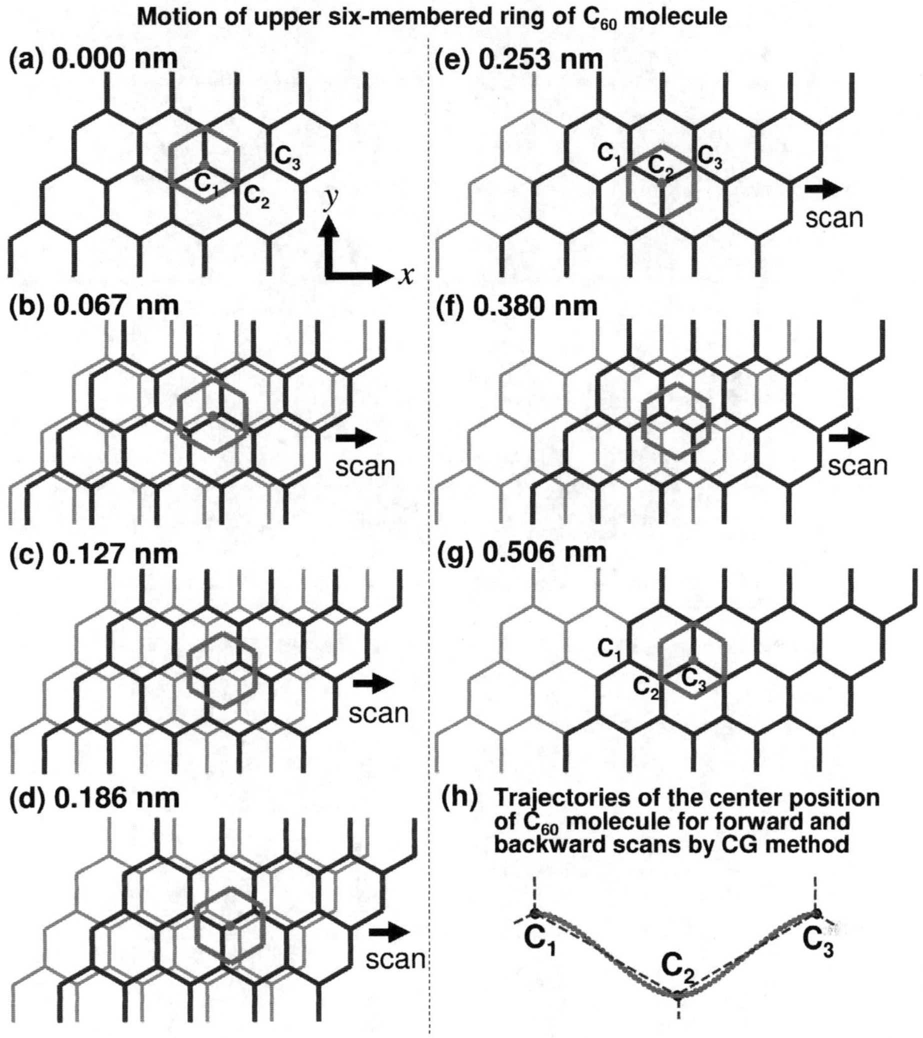

Figure 4. During the scanning process of the blue upper graphene sheet, red upper six-membered ring of the C_{60} molecule slides nearly along the green carbon bond of the lower graphene sheet or the blue carbon bond of the upper graphene sheet. The red center position of the six-membered ring is located nearly on the carbon bonds of the lower and upper graphene sheets, for the scan position $L =$ (a) 0, (b) 0.067, (c) 0.127, (d) 0.186, (e) 0.253, (f) 0.380, and (g) 0.506 nm. (h) Trajectories of the center position of the C_{60} molecule on the carbon bonds C_1-C_2-C_3 of the green lower graphene lattice for the forward (red) and backward (blue) scans for 0 nm $\leq L \leq$ 0.506 nm. The trajectories for the forward and backward scans are completely the same as each other. The C_{60} molecule takes a sinusoidal motion along the zigzag carbon bonds C_1-C_2-C_3.

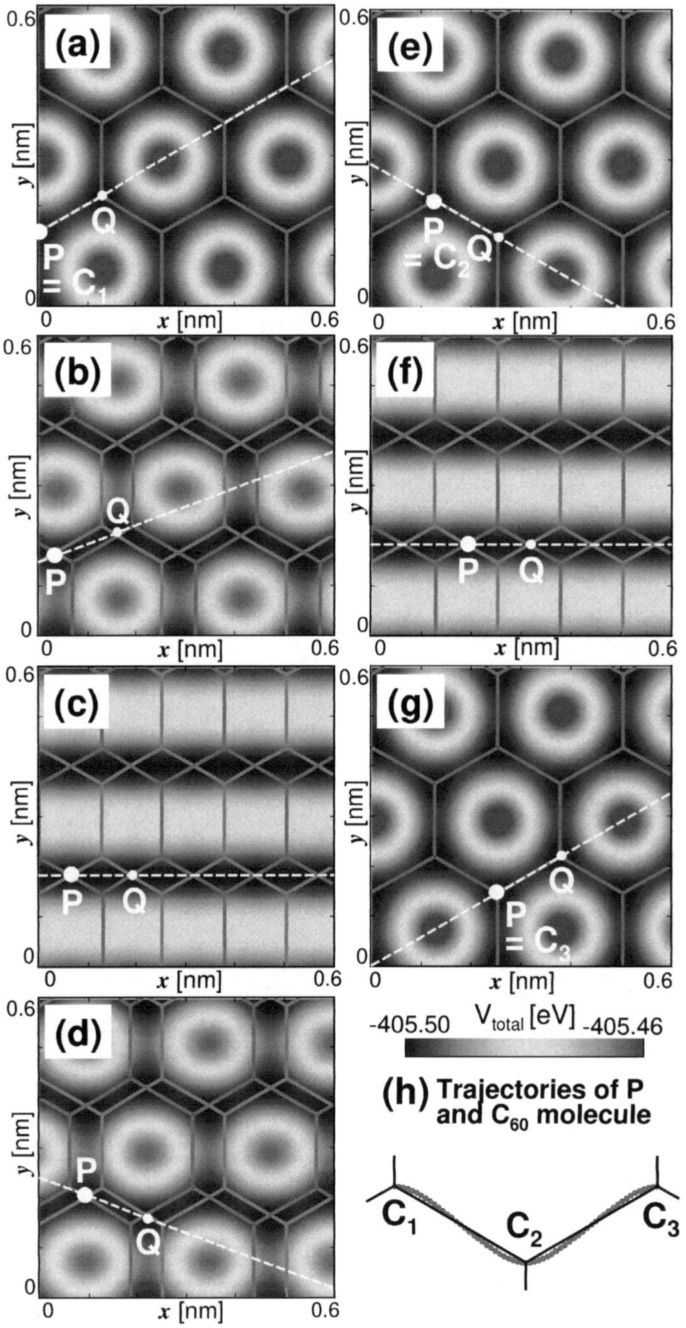

Figure 5. Total energy V_{total} plotted as a function of the center position of the C_{60} molecule (x, y) for the scan position of the upper graphene sheet, $L =$ (a) 0, (b) 0.067, (c) 0.127, (d) 0.186, (e) 0.253, (f) 0.380, and (g) 0.506 nm. Red honeycomb network (upper graphene sheet) is scanned and the violet one (lower graphene sheet) is fixed. Circles P and Q denote the nearest neighboring local minima. The larger circle P corresponds to the center position of the C_{60} molecule. The cross sections along the broken line connecting P and Q are illustrated in Fig. 6. (h) Trajectories of P obtained by directly checking the total potential energy V_{total}, and the center position of the C_{60} molecule calculated by the CG method, which are completely the same as each other.

red round regions as shown in Fig. 5(a). Furthermore the saddle points of $V_{\rm total}$ appear at the center position of the carbon bond.

As L is varied, the two-dimensional pattern of $V_{\rm total}\,(x,y;L)$ markedly changes as follows. For $L=0.067$ nm, the minimum position P moves nearly along the carbon bond and the red regions surrounding the maximum points extend along the scan direction to become elliptic shape as shown in Fig. 5(b). Then, for $L=0.127$ nm, half of the period along $[12\bar{3}0](=x)$ direction, P is located on the center position of the carbon bond as shown in Fig. 5(c). Here the red elliptic regions vanish, which results in the formation of the green and blue stripe patterns (Fig. 5(c)). For $L=0.186$ nm, P moves further nearly along the carbon bond, and the red regions take elliptic shape the same as that for $L=0.067$ nm (Fig. 5(b)) again as shown in Fig. 5(d). Then, as shown in Fig. 5(e), for $L=0.253$ nm, the red round regions just the same as the initial one for $L=0$ nm (Fig. 5(a)) appear again. Here P is located on the carbon atom C_2. For $0.253 \leq L \leq 0.506$ nm (Figs. 5(e)-5(g)), the movement of the minimum position P and the variation of the two-dimensional patterns of $V_{\rm total}\,(x,y;L)$ are just the opposite to those for $0 \leq L \leq 0.253$ nm. Thus the initial minimum position P for $L=0$ nm becomes completely equivalent to P for $L=0.506$ nm.

Fig. 5(h) shows two kinds of trajectories. One is the point P obtained by directly checking the local minimum position located on the total potential energy surface $V_{\rm total}$. The other is the center position of the C_{60} molecule obtained by the structural optimization, CG method, which is the same as Fig. 4(h) for the forward scan. As shown in Fig. 5(h), the minimum position P and the center position of the C_{60} molecule are completely the same as each other. This means analysis of the total potential energy gives the same results of the C_{60} motion as the CG method.

3.4. Energy barrier

Figs. 6(a)-6(g) shows the cross sections along the yellow-dotted line in Figs. 5(a)-5(g) connecting the nearest neighboring minimum positions, P and Q. During all the scan processes, there exists an energy barrier between P and Q, which prevents the C_{60} molecule from jumping from P to Q under $T \to 0$ K from the static point of view. Therefore the C_{60} molecule can continuously slide kept to be located on the local minimum position P. Fig. 6(h) shows the activation energy barrier $\Delta V_{\rm total}$ between P and Q plotted as a function of the scan position of the upper graphene sheet, L. $\Delta V_{\rm total}$ takes the maximum value of 3.8 meV for $L=0.245$ and 0.498 nm, and takes the minimum value of 0.83-0.84 meV for $L=0.110$ and 0.371 nm. At the room temperature, even the maximum energy barrier of 3.8 meV is low enough to be jumped due to the thermal activation where various types of translation, rotation and oscillation of the C_{60} molecule are expected to occur easily. Nevertheless the effect of the energy barrier of 3.8 meV on the C_{60} dynamics clearly appears under the ultralow temperature condition of $T \to 0$ K.

4. Conclusions and Discussions

In this paper we numerically studied the unique near-zero friction regime along the $[12\bar{3}0]$ direction of the C_{60} molecular bearings. The mechanism of the conservative and reversible motion of the C_{60} molecule can be understood by the structural optimization based on the CG method and direct analysis of the minimum position of the total potential energy $V_{\rm total}$ as follows:

(1) The lateral force curve shows the sinusoidal behavior with a period of the graphite lattice along $[12\bar{3}0]$ direction, 0.253 nm. The lateral force curves are completely the same as each other between the forward and backward scans.

(2) The evaluation of the tilting angles of the C_{60} molecule gives us information that, for the interlayer distance $d=1.3$ nm, the C_{60} molecule slides with its six-membered ring faced nearly parallel to both the upper and lower graphene sheets during the scan.

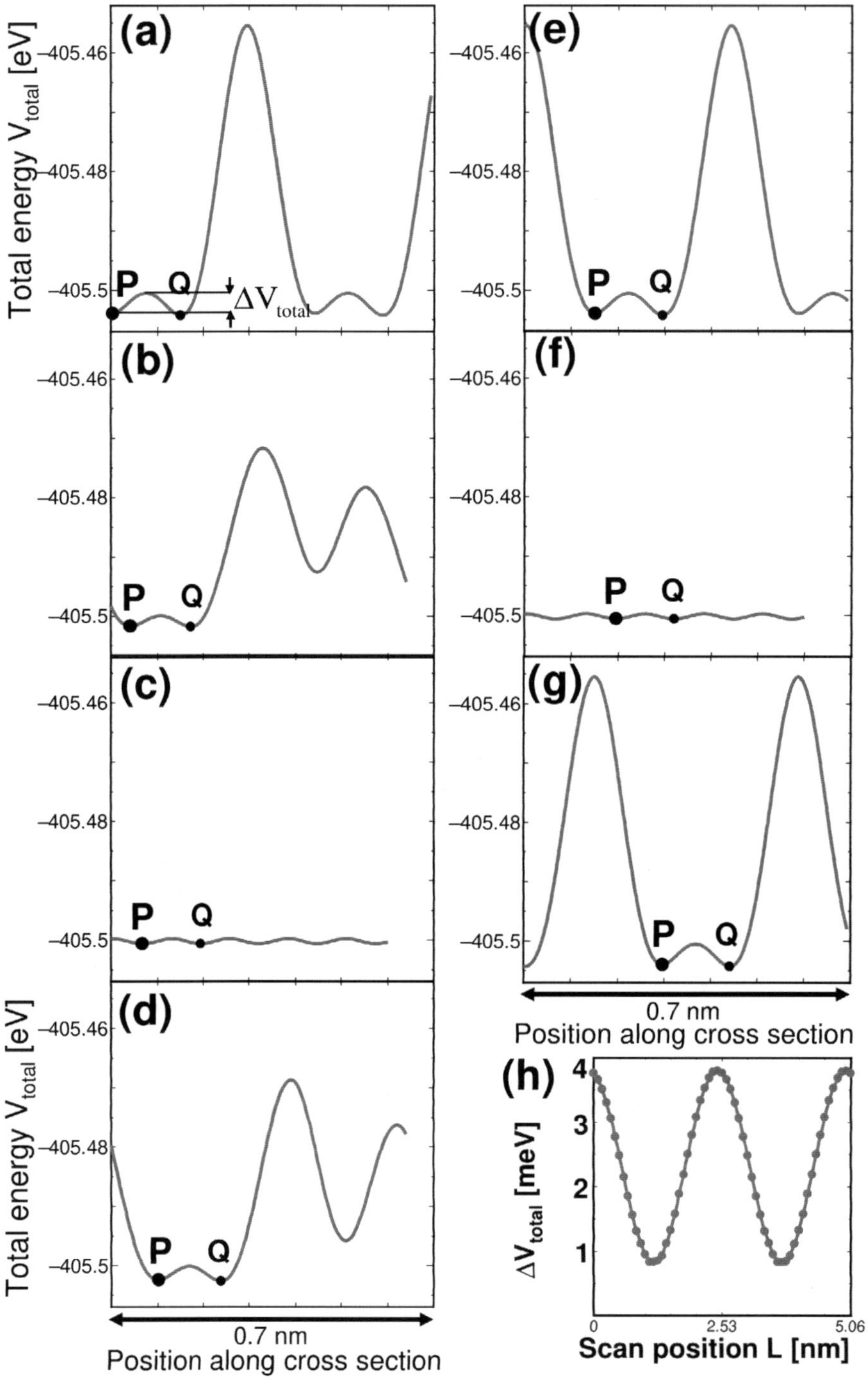

Figure 6. The cross section of the total energy surface V_{total} along the broken line connecting the minimum positions P (center position of the C_{60} molecule) and Q (nearest neighboring minimum to P), for $L =$ (a) 0, (b) 0.067, (c) 0.127, (d) 0.186, (e) 0.253, (f) 0.380, and (g) 0.506 nm. (h) The energy barrier between the neighboring minima P and Q, $\Delta V_{\mathrm{total}}$, illustrated in Figs. 6(a)-6(g), plotted as a function of L, the scan position of the upper graphene sheet.

(3) The six-membered ring of the C_{60} molecule slides along the path $C_1 \rightarrow C_2 \rightarrow C_3$ for the forward scan along the $[12\bar{3}0]$ $(=x)$ direction within $0 \leq L \leq 0.506$ nm. For the backward scan along $-x$ direction, the six-membered ring of the C_{60} molecule takes a sliding motion $C_3 \rightarrow C_2 \rightarrow C_1$ just opposite to that for the forward scan. The trajectories of the C_{60} molecule with a sinusoidal shape for the forward and backward scans are completely the same as each other. This means the C_{60} motion is reversible and continuous for the forward and backward scans of the upper graphene sheet along $[12\bar{3}0]$ $(=x)$ direction. It should be noted that the physical origin of the sinusoidal shape of the trajectory can be understood by considering the shape of the energy potential surface V_{total} for each L.

(4) The C_{60} motion mentioned in (3) can be understood by calculating the minimum position P of the total potential energy surface $V_{\mathrm{total}}\,(x, y; L)$ during the scanning process. It is confirmed that the trajectories of P obtained by directly checking V_{total}, and the center position of the C_{60} molecule obtained by the structural optimization, CG method, are completely the same as each other.

(5) It is clarified that, during all the scanning processes, there exists an energy barrier between the nearest neighboring minimum positions, which prevents the C_{60} molecule from jumping toward the nearest neighboring minimum position under $T \rightarrow 0$ K from the static point of view. Therefore C_{60} molecule can slide continuously kept to be located on the local minimum position.

In this paper, we discuss the low loading condition corresponding to the graphene interlayer distance of 1.3 nm. For the smaller interlayer distance corresponding to the higher loading condition, effect of the C_{60} tilting or rotation can give marked influences on the superlubricity similar to the case of the scan along the $[10\bar{1}0]$ direction [7, 8, 9]. Therefore the superlubricity of the C_{60} molecular bearings, graphene/C_{60}/graphene interface, for the $[12\bar{3}0]$ scan under the low loading and low temperature conditions is very unique. If this system including the loading and temperature conditions is realized, there is a possibility of energy-conserved transportation of the nanoscale objects by using the stacking registry of the carbon-network. This condition can be achieved more easily in the system of graphene/graphene/graphene stacking interface, which will be reported elsewhere.

Acknowledgments
This research was supported by a Grant-in-Aid for Scientific Research (B) (No. 20360022) and Specially Promoted Research (No. 21000008) from the Japan Society for the Promotion of Science. This work was also supported by the Ministry of Education, Culture, Sports, Science and Technology through a Grant-in-Aid for Building Strategic Research Infrastructures.

References
[1] Miura K, Kamiya S and Sasaki N 2003 *Phys. Rev. Lett.* **90** 055509
[2] Sasaki N and Miura K 2004 *Jpn. J. Appl. Phys.* **43** 4486
[3] Miura K, Tsuda D and Sasaki N 2005 *e-J. Surf. Sci. Nanotech.* **3** 21
[4] Miura K, Tsuda D and Sasaki N 2005 *Tribologist* **49** 553
[5] Miura K, Tsuda D, Itamura N and Sasaki N 2007 *Jpn. J. Appl. Phys.* **46** 5269
[6] Itamura N, Miura K and Sasaki N 2009 *Jpn. J. of Appl. Phys.* **48** 060207
[7] Sasaki N, Itamura N and Miura K 2007 *J. Phys.: Conf. Ser.* **89** 012001
[8] Sasaki N, Itamura N and Miura K 2007 *Jpn. J. of Appl. Phys.* **46** L1237
[9] Itamura N, Miura K and Sasaki N 2009 *Jpn. J. of Appl. Phys.* **48** 030214
[10] Miura K and Kamiya S 2002 *Europhys. Lett.* **58** 610
[11] Tersoff J 1988 *Phys. Rev. Lett.* **61** 2879
[12] Lu Jian Ping, Li X -P and Martin Richard M 1992 *Phys. Rev. Lett.* **68** 1551
[13] Stoddard S D and Ford J 1973 *Phys. Rev.* **A8** 1504
[14] Press W H, Teukolsky S A, Vetterling W T and Flannery B P 1992 *Numerical Recipes: The Art of Scientific Computing*, 2nd ed (New York: Cambridge Univ. Press) pp 413-418

Journal of Physics: Conference Series **258** (2010) 012014 doi:10.1088/1742-6596/258/1/012014

Evaluation of friction transition for metal-semiconductor interfaces using model potential comprising three-body contributions

Motohisa Hirano[1], Hiroki Murase[1], Takahiro Nitta[1], and Tomonori Ito[2]

[1]Department of Mathematical and Design Engineering, Gifu University,1-1 Yanagido, Gifu, 501-1193, Japan

[2]Department of Physics Engineering, Mie University, 1577 Kurima-Machiya, Tsu 514-8507, Japan

E-mail: hirano@gifu-u.ac.jp

Abstract.
Whether or not the friction transition [1, 2] occurs in the frictional systems of W(011) and Si(001) atomically clean surfaces has been examined in relation to a previous ultrahigh vacuum scanning tunneling microscopy experiment investigating friction transition. This examination takes into account of empirical inter-atomic potentials with three-body interactions. To obtain equilibrium atomic arrangements of the frictional system for evaluating friction transition, the parameters of the interatomic potential applicable for tetrahedrally bonded materials were examined. From studying the criterion of the friction transition for the frictional systems of W(011) and Si(001), it has been concluded that friction transition does not occur for the real systems, which supports the experimental results of a previous ultrahigh vacuum scanning tunneling microscopy experiment.

1. Introduction

The total energy of contacting surfaces is invariant for sliding distance Q as in the case of incommensurate interfaces, resulting in vanishing frictional force. The increase in interfacial interactions, on the other hand, leads to the discontinuous motion of constituent atoms, resulting in the appearance of friction. We thus see that the frictional force changes from vanishing to finite due to increased interfacial interactions. This is called friction transition [1]. Whether the friction transition occurs in real systems is unknown. The occurrence of the friction transition has been examined by deriving its condition in various inter-metallic systems by using pair-wise inter-atomic potentials [1, 3]. A related experiment has been conducted using a scanning tunneling microscopy technique in the frictional systems of W(011) tips and Si(001) atomically clean surfaces. The conclusion was that the friction transition does not occur in a frictional system [4, 5]. This paper, in relation to that experiment, theoretically examines whether or not friction transition occurs in the frictional systems of W(011) and Si(001) atomically clean surfaces, by taking into account empirical inter-atomic potentials with three-body interaction [6].

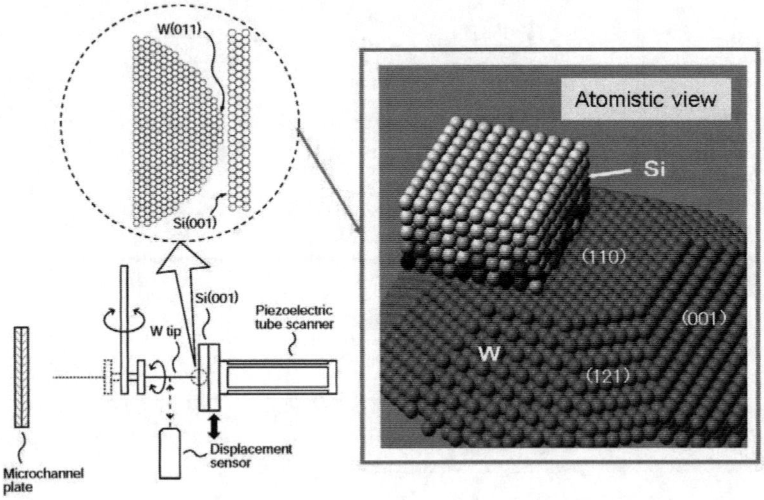

Figure 1. Schematic illustration of STM friction measurement system in an ultra-high vacuum with base pressure of 10^{-9} Pa. Cleanliness and crystal orientation of W(011) surface of tungsten tip were determined by a field emission microscope, and that of Si(001) surface by low energy electron diffraction and Auger electron spectroscopy.

2. Experiments for W(001) and Si(001) interfaces in ultrahigh vacuum

Ultrahigh vacuum scanning tunneling microscopy (STM) technology has been used to study friction between atomically clean, well-defined crystal surfaces equivalent to theoretical models [4]. In an ultra-high vacuum (10^{-9} Pa), the W(011) surface of a tungsten (W) tip is brought close to a Si(001) surface (Sb-doped, n-type, 0.01 Ω·cm) (Fig. 1). Then, while the tunnel gap between the two surfaces is kept constant, the Si(001) surface is one-dimensionally scanned by the tip and the friction force between the surfaces is measured from the deflection of the tip at the time of scanning (Fig. 1). Quantum tunneling was used to prevent plastic deformation caused by contact between the tip and the surface. In contrast to ordinary STM, the tungsten tip is not sharpened on purpose and the minute surface, as shown in Fig. 1, prepared on the tip is used as a friction surface.

Experimental results agreed with theoretical predictions, and a friction force equivalent to the theoretical one was observed for commensurate contact [4]. A friction force of 0.8 nN was obtained for a tip elastic modulus of 1.5 N/m and tip deflection of 50 nm in agreement with theoretical values. On the other hand, friction force could not be observed for incommensurate contact at an observation accuracy of 3 nN. In the above manner, the experimental results for clean surfaces at the atomic level agreed with the theoretical prediction. It was concluded that friction forces equivalent to theoretic values can be observed for commensurate contact while friction forces cannot be observed for incommensurate contact at a measurement resolution of 3 nN.

3. Criterion for friction transition

According to Tomlinson's mechanism[2], atoms at contacting surfaces discontinuously (abruptly) change their positions when solids slide against each other. The occurrence of the discontinuity means that the atoms can not take the arbitrary equilibrium position. This is equivalent to a condition that the second derivatives of the total potential energy $V(Q, r)$ with respect to

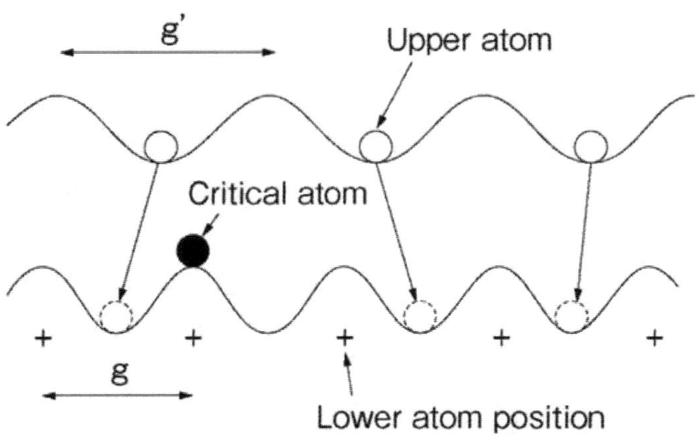

Figure 2. One-dimensional frictional system. Predictions of $V^l(r)$ and $V^u(r)$ characterized by lengths g and g', respectively. Atoms sit on lowest minima of $V^u(r)$, when $V^l(r) = 0$. When $V^l(r)$ becomes strong limit, atoms occupy positions for lowest minima of $V^l(r)$.

the coordinates of the atom r becomes be negative at the position where $d^2V^l(r)/dr^2$ takes the largest negative value (see Fig. 2). A simple case involving a one-dimensional system is shown, in which only $V^l(r)$ and $V^u(r)$ operate (Fig. 2). When $V^l(r)$=0, the atoms occupy positions that correspond to the lowest minima of $V^u(r)$. For a weak $V^l(r)$ limit, the atoms slightly change their positions towards the minimum positions of $V^l(r)$. For a strong $V^l(r)$ limit, the atoms occupy positions that correspond to the lowest minima of $V^l(r)$. In the intermediate case, does the critical atom, shown in Fig. 2, sits an arbitrary position without showing the discontinuous motion of atoms? The criterion for the occurrence of the discontinuous motion is given in Ref. [1] as

$$\frac{d^2V(Q, r_{max})}{dr^2_{max}} = \frac{d^2V^u(Q - r_{max})}{dr^2_{max}} + \frac{d^2V^l(r_{max})}{dr^2_{max}} < 0. \quad \text{for a certain } Q, \tag{1}$$

where $V^l(r)$ is the interaction from the atoms of the lower body. $V^u(r)$ is the interaction from the atoms of the second, third, ... layers of the upper body. r_{max} stands for the position defining the largest negative value of $d^2V^l(r)/dr^2$, and Q stands for the displacement coordinate of the sliding upper body against the lower one.

Note that there is the case where discontinuous motion does not occur. This result suggests that the discontinuous motion is likely to occur only when the interaction or adhesion between two bodies becomes strong and successive discommensuration [7] occurs. This suggestion is generally valid since the equilibrium positions of atoms of the upper body are slightly disturbed by the lower body for the weak case of adhesion [8, 9]. It has been thus shown that the frictional force changes from vanishing to finite due to increased interfacial interactions in incommensurate systems. This is called **"friction transition"** [1]. Friction transition is the same as the transition of analyticity-breaking, often called Aubry transition [7]. Aubry studied the Frenkel-Kontorova model [10], which is a one-dimensional system to describe the movement of defects or dislocations. Analyticity-breaking corresponds to the discontinuous change of the relaxed particle positions.

The domains where the atom can take its equilibrium position when the adhesion increases between the upper and lower bodies [5] are shown in Fig. 3. The difference from the one-

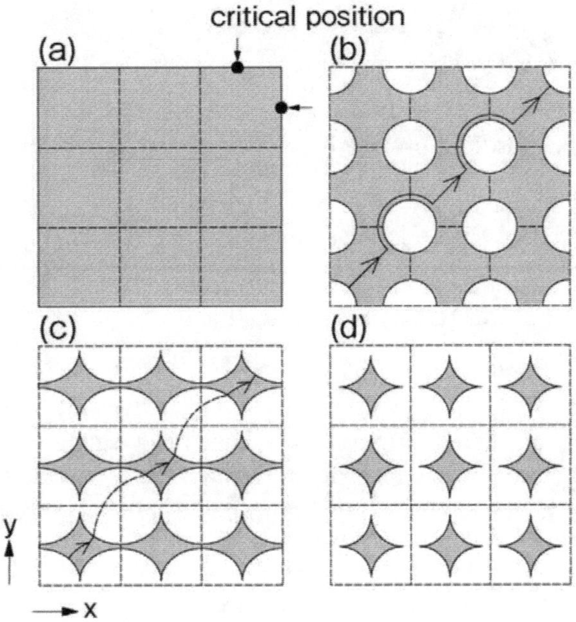

Figure 3. Topological property of the whole domain, where the atom can take its equilibrium position. Figures show the cases where the contacting surface of the lower body has the oblique-square crystalline symmetry.

dimensional case is that the path where the atom can slide by continuously changing its equilibrium position depends on the direction of sliding displacement coordinate Q. The atom can slide towards an arbitrary direction by continuously changing its equilibrium position, as in the case shown in Figs. 3(a) and (b). In the case shown in Fig. 3(c), the atom can slide continuously in the x-direction, but can slide in the y-direction only by discontinuous transformation. The atom can slide only by discontinuously changing its equilibrium position in any direction for the case shown in Figure 3(d). As seen in these figures, the frictional property depends on the topological property, i.e., the disconnectedness or connectedness of the domains where the atom can take the equilibrium position.

4. Friction transition for metal-semiconductor interfaces

The main concern here is whether or not friction transition occurs in the frictional systems of W(011) and Si(001) atomically clean surfaces, which were measured by a ultrahigh vacuum STM technology [4]. It is necessary to assess the interatomic potential for tetrahedrally bonded materials. The empirical interatomic potentials proposed by Khor and Das Sarma [6] have been adopted for evaluating the frictional system. This potential, which uses very few parameters, is useful, particularly for surface studies, including the surface reconstruction of Si surfaces. The potential is given by

$$V_{ij} = A\exp[-\beta(r_{ij} - R_i)^{\gamma}]\left[\exp(-\theta r_{ij}) - \frac{B_0}{Z_i^{\alpha}}\exp(-\lambda r_{ij})G(\eta)\right], \tag{2}$$

where r_{ij} is the interatomic distance, R_i is the minimum interatomic distance of the neighbors, and $A, B_0, \theta, \beta, \gamma$, and λ are respectively constants. The effective coordination number of atoms is

$$Z_i = \sum_j \exp[-\beta(r_{ij} - R_i)^\gamma], \tag{3}$$

where the summation is carried over all neighbors of i. The parameters β and γ can be fitted to give the correct effective coordination numbers of crystal structures of materials. $G(\eta)$ is given by

$$G(\eta) = 1 + \sum_{k \neq i,j} [\cos(\eta \Delta\theta_{jik}) - 1], \tag{4}$$

where η is a parameter for the bond-bending force constant, and the summation is carried out over the nearest-neighbor atoms. $\Delta\theta_{jik}$ is expressed as

$$\Delta\theta_{jik} = |\theta_{jik} - \theta_i|, \tag{5}$$

where θ_i is the equilibrium angle between nearest-neighbor bonds for a regular structure with the coordination number Z_i (i.e., 109.47° for a diamond structure). θ_{jik} is the angle between bonds ij and ik, η is a parameter to be fitted to the bond-bending force constant, and the summation is to be carried out only over nearest-neighbor atoms whose bonds ik are the nearest to ij. θ_i is given by

$$\theta_i = \frac{2}{3}\pi + A_2 x^2 + A_3 x^3 + A_4 x^4 + A_5 x^5, \tag{6}$$

$$x = \ln(Z_i) - \ln(3), \tag{7}$$

where $A_2, A_3, A_4, and A_5$ are parameters given by A_2=-4.137495128948, A_3=8.5530372320793, A_4=-7.1599649270085, and A_5=2.0627878449403 respectively.

In Table 1, we give the values of the parameters of Eqs.(2)-(4) for the three pairs Si, W, and W-Si, respectively. These are fitted to the measured bulk modulus and the cohesive energy shown in Table 2.

Table 1. Potential parameter values for Si, W, and W-Si pairs.

Parameters	Si	W	W-Si
A	2794.2386	3798.6189	38444.1675
B_0	0.08251716	0.251094	0.169645
θ	3.13269	2.68935	2.79434
λ	1.34146	1.52280	1.70172
α	0.6249096	0.548241	0.285641
β	25.44123	20.75723	25.42105
γ	3.38218	3.26327	3.39927
η	0.90084597	0	0

The friction transition calculation follows the same procedures as described elsewhere [1]. Figure 4 shows the atomistic model of Si(001) and W(011) for evaluating friction transition. The upper body shows the Si diamond structure, and the lower body shows the W face-centered tetragonal lattice.

It is assumed that only the atoms of the upper body are allowed to change their positions while the atoms of the lower body remain fixed. To satisfy this assumption, the closed packed

Table 2. Experimental data for determining potential parameters. r_e is equilibrium interatomic spacing, E_{coh} is cohesive energy, and B is bulk modulus.

Material	Structure	r_e(Å)	E_{coh}(eV/atom)	$B(10^{11}\mathrm{N/m^2})$
W	bcc	2.74	8.90	3.232
	fcc	2.84	8.414	—
W-Si	Zinc blende	2.44	5.308	1.087
	Rocksalt	2.55	6.214	—
	CsCl	2.68	6.667	—

crystal planes (hard planes), such as the W(011) planes for tungsten lattices, are taken as the contact surfaces of the lower bodies. The atoms of the upper body Si lattice were relaxed by operating the interatomic force given by Eqs.(8)-(12).

$$\begin{aligned}\frac{\partial V_{ij}}{\partial r_{ij}} &= -A\beta\gamma(r_{ij}-R_i)^{\gamma-1}\exp[-\beta(r_{ij}-R_i)^{\gamma}]\left[\exp(-\theta r_{ij}) - B_0\exp(-\lambda r_{ij})\frac{G(\eta)}{Z_i^{\alpha}}\right] \\ &+A\exp[-\beta(r_{ij}-R_i)^{\gamma}]\left[-\theta\exp(-\theta r_{ij}) - \left\{-\frac{B_0\alpha}{Z_i^{\alpha+1}}\nabla_j(Z_i)\exp(-\lambda r_{ij})G(\eta)\right.\right. \\ &\left.\left.-\frac{B_0}{Z_i^{\alpha}}\lambda\exp(-\lambda r_{ij})G(\eta) + \frac{B_0}{Z_i^{\alpha}}\exp(-\lambda r_{ij})\nabla_j\{G(\eta)\}\right\}\right]. \end{aligned} \quad (8)$$

$$\nabla_j(Z_i) = \sum_j -\beta\gamma(r_{ij}-R_i)^{\gamma-1}\exp\{-\beta(r_{ij}-R_i)^{\gamma}\}. \quad (9)$$

$$\nabla_j\{G(\eta)\} = \sum_k \frac{\eta\sin(\eta\Delta\theta_{jik})}{\sin\theta_{ijk}}\nabla_j(\cos\theta_{ijk}) + \eta\sin(\eta\Delta\theta_{jik})\nabla_j(Z_i). \quad (10)$$

$$\nabla_j(\cos\theta_{ijk}) = \frac{1}{r_{ij}}(\hat{r}_{ik} - \hat{r}_{ij}\cos\theta_{ijk}). \quad (11)$$

$$\nabla_j(\theta_i) = (2A_2x + 3A_3x^2 + 4A_4x^4 + 5A_5x^5)\frac{1}{Z_i}\nabla_j(Z_i), \quad (12)$$

The Si(001) plane is places against the W(011) plane, as shown in Fig. 4. The upper body is composed of $5\times5\times5$ diamond structure unit cells (1166 atoms), and the lower body is composed of $18\times18\times2$ face-centered tetragonal unit cells (3423 atoms). The corresponding interfaces with misfit angles of 0 and 30° respectively are shown in Fig. 5

The critical atom for each contact is placed at the point satisfying $\partial V^l(r)/\partial r = 0$ and $\partial^2 V^l(r)/\partial r^2 < 0$ along lines perpendicular to the ridge lines of $V^l(r)$, where the local minimum of potential $V^u(r_{min})$ from the upper body and the local maximum of potential $V^l(r_{max})$ from the lower body exist. In the calculation, the atoms of the upper body, excluding the critical atoms, are three-dimensionally relaxed. After relaxation, second order derivative $V_{\alpha,\beta}$ is calculated for the direction perpendicular to the ridge line of each critical atom.

Figure 6 shows the calculated $V_{\alpha,\beta}$ as a function of the misfit angle at the contacting surfaces. The calculated $V_{\alpha,\beta}$ values are positive for the misfit angle between 0 and 90°. This shows that friction transition does not occur in the frictional system of Si(001) and W(011) surfaces. The $V_{\alpha,\beta}$ sign is actually determined by competition between a positive $V^u(r_{min})$ contribution from the upper body and a negative $V^l(r_{max})$ contribution from the lower body as seen in Eq. (1). A positive $V^u(r_{min})$ always defeats a negative $V^l(r_{max})$ in the frictional system. From studying the frictional properties of Si(001) and W(011) surfaces, it has been concluded that friction transition does not occur for real systems, which supports the experimental results of the ultrahigh vacuum STM experiment [4].

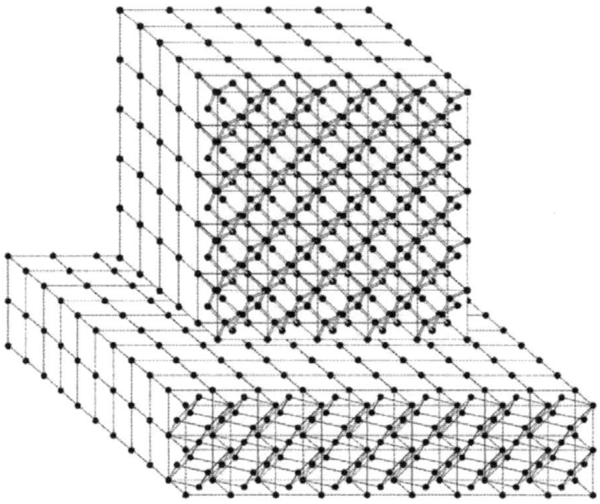

Figure 4. Atomistic model of Si(001) and W(011) for evaluating friction transition.

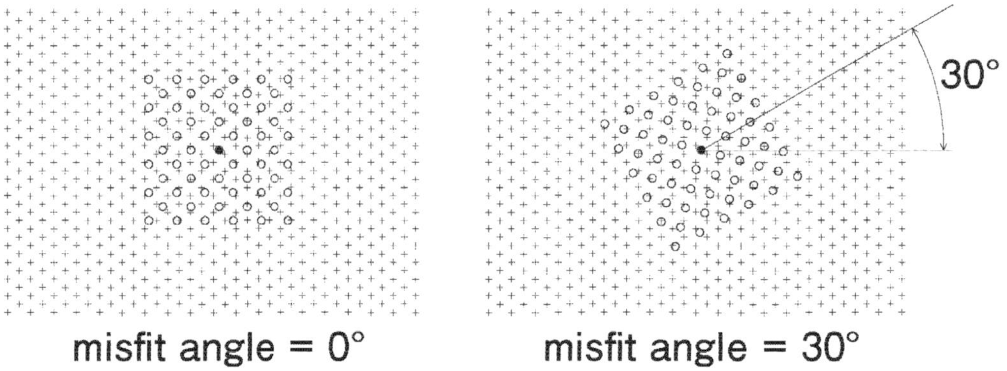

Figure 5. Interface model for friction transition in realistic systems of Si(001) and W(011).

5. Summary

Whether or not the friction transition occurs has been examined by studying the atomistic frictional systems of W(011) and Si(001). The interatomic potential, comprising three-body contributions, for tetrahedrally bonded material was used. The construction of the potential is based on the idea that bonding energies of many substances can be modeled by pairwise interactions moderated by the local environment. The friction transition was calculated by following the same procedures as described elsewhere [1]. From studying the frictional properties of Si(001) and W(011) surfaces, which were actually measured by STM techniques, the calculated $V_{\alpha,\beta}$ values were found to be all positive for various contacting conditions of W(011) and Si(001). This proves that friction transition does not occur for real systems, which supports the experimental results of the ultrahigh vacuum STM experiment [4]. Friction transition does not occur even for strong adhesion such as metallic bonding [1]. The case of weak adhesion has been discussed elsewhere [8, 9]. Thus, the present conclusion may hold true for a wider class of

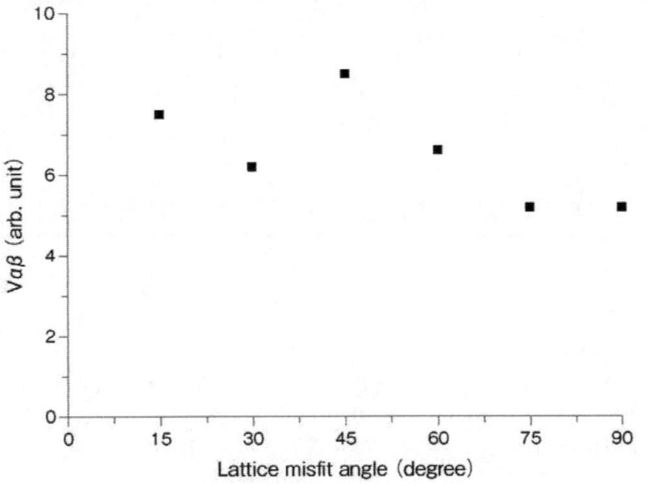

Figure 6. Calculated $V_{\alpha,\beta}$ as function of misfit angle θ of contacting surfaces.

adhesion.

References

[1] Hirano M and Shinjo K 1990 *Phys. Rev. B* **41** 11837
[2] Tomlinson G A 1929 *Philo. Mag.* **7** 905
[3] Girifalco L A and Wezer V G 1959 *Phys. Rev.* **114** 687
[4] M Hirano K Shinjo R K and Murata Y 1997 *Phys. Rev. Lett* **78** 1448
[5] Hirano M 1990 *Surface Science Reports* **60** 159
[6] Khor K E and Sarma S D 1990 *Phys. Rev. B* **38** 3318
[7] Aubry S 1983 *J. Physique* **44** 147
[8] McClelland G M 1990 *Adhesion and Friction Springer Series in Surface Science* vol 17 ed Grunze M and Kreuzer H J (Berlin: Springer Verlag)
[9] Sokoloff J B 1984 *Surf. Sci.* **144** 267
[10] Frenkel Y I and Kontorowa T 1938 *Zh. Eksp. Teor. Fiz.* **8** 1340

Journal of Physics: Conference Series **258** (2010) 012015 doi:10.1088/1742-6596/258/1/012015

Dissipation and entropy production in open quantum systems

H Majima, A Suzuki

Department of Physics, Faculty of Science, Tokyo University of Science, 1-3 Kagurazaka, Shinjuku-ku, Tokyo 162-8601 Japan.

E-mail: majima@rs.kagu.tus.ac.jp, asuzuki@rs.kagu.tus.ac.jp

Abstract. A microscopic description of an open system is generally expressed by the Hamiltonian of the form: $H_{\rm tot} = H_{\rm sys} + H_{\rm environ} + H_{\rm sys-environ}$. We developed a microscopic theory of entropy and derived a general formula, so-called "entropy-Hamiltonian relation" (EHR), that connects the entropy of the system to the interaction Hamiltonian represented by $H_{\rm sys-environ}$ for a nonequilibrium open quantum system. To derive the EHR formula, we mapped the open quantum system to the representation space of the Liouville-space formulation or thermo field dynamics (TFD), and thus worked on the representation space $\mathscr{L} := \mathscr{H} \otimes \widetilde{\mathscr{H}}$, where $\mathscr{H}$ denotes the ordinary Hilbert space while $\widetilde{\mathscr{H}}$ the tilde Hilbert space conjugates to $\mathscr{H}$. We show that the natural transformation (mapping) of nonequilibrium open quantum systems is accomplished within the theoretical structure of TFD. By using the obtained EHR formula, we also derived the equation of motion for the distribution function of the system. We demonstrated that by knowing the microscopic description of the interaction, namely, the specific form of $H_{\rm sys-environ}$ on the representation space $\mathscr{L}$, the EHR formulas enable us to evaluate the entropy of the system and to gain some information about entropy for nonequilibrium open quantum systems.

1. Introduction

Dynamical processes in dissipative (open) systems involve irreversible transformations of nonequilibrium states of the systems. An important physical quantity of the theory of non-equilibrium processes is the entropy production that is the rate at which entropy is produced as a result of irreversible processes [1, 2]. In the quantum theory of open systems the entropy production is usually related to the negative derivative of the relative entropy with respect to an invariant state in order to show the second law of thermodynamics [1]. However, conventional theories have not made clear the connection between the entropy production and the interactions between the system and its environment in open quantum systems. Indeed providing a microscopic expression for the entropy production has been one of the grand aims of statistical mechanics, going back to the seminal work of Boltzmann.

In this paper, we derive a general formula which relates the entropy of the system to the system-environment interactions in an open quantum system (generally represented by the Hamiltonian $H_{\rm tot} = H_{\rm sys} + H_{\rm environ} + H_{\rm sys-environ}$). To derive the formula, we adopt the Liouville-space formulation or the method of thermo field dynamics (TFD) [3, 4] as a means for treating open quantum systems. In the formulation of the problem, we shall use the idea introduced in the theory of TFD, namely, doubling the degree of freedom by introducing the tilde Hilbert space $\widetilde{\mathscr{H}}$, and the rules defined in the representation spaces $\widetilde{\mathscr{H}}$ and $\mathscr{H}$ introduced in TFD. In concreteness, we map the open quantum system to the representation space of TFD. On the basis of the zeroth law of thermodynamics [5, 6, 7], it can be shown that the interactions between a system and its environment can be transformed into the nontilde-tilde interactions [8, 9]. Once the entropy of the system is expressed on the Hilbert space $\mathscr{H}$ and is connected to the system-environment interaction expressed on the representation space $\mathscr{L} := \mathscr{H} \otimes \widetilde{\mathscr{H}}$, we can examine the behavior of the entropy from the microscopic description of the interaction Hamiltonian $H_{\text{sys}-\text{environ}}$. In other words, we can study the behavior of the entropy of the system microscopically from the knowledge of those interactions between the system represented on the space $\mathscr{H}$ and its environment represented on the space $\widetilde{\mathscr{H}}$ within the framework of TFD. By studying the interactions between the system and its environment mapped on the representation space $\mathscr{L}$, we can generally derive a formula which connects the interaction Hamiltonian to the entropy of the system in open quantum systems.

We also show that the equation of motion for the distribution function of the system can be derived from the formula and it is shown that the entropy of the system is expressed in terms of the distribution function. Finally, we consider two possible example models as applications of the formulas obtained for the open quantum systems.

2. Entropy-Hamiltonian relation

In this section, we discuss an open quantum system in connection with the framework of TFD and derive the formula which connects the entropy and the Hamiltonian (so called *entropy-Hamiltonian relation*, EHR for short) for open quantum systems by using an entropy operator and the generalized Schrödinger-like equation in the representation space $\mathscr{L}$.

Let us briefly explain TFD. TFD is constructed by doubling a representation space for a set of operators $\{A\}$ defined on the Hilbert space $\mathscr{H}$ by introducing the tilde conjugation Hilbert space $\widetilde{\mathscr{H}}$ for a set of those tilde conjugate operators $\{\widetilde{A}\}$. Utilizing this structure in the theory of TFD, we can conveniently express the interaction Hamiltonian of an open quantum system in terms of those operators on $\mathscr{H}$ and on $\widetilde{\mathscr{H}}$ in the representation space $\mathscr{L}$ as will be shown shortly. Bosonic and fermionic tilde operators are, respectively, assumed to commute and to anticommute with respective nontilde operators.

Each operator in $\{\widetilde{A}\}$ satisfies the tilde conjugation rules [4]. Thermal state $|\Psi(t)\rangle\rangle$ defined in TFD is expressed by

$$|\Psi_\alpha(t)\rangle\rangle = \rho^\alpha(t)|1\rangle\rangle \qquad (0 \le \alpha \le 1), \tag{1}$$

where $\rho(t)$ is a density operator describing a system, and a state $|1\rangle\rangle$ is defined by $|1\rangle\rangle := e^{a^\dagger \tilde{a}^\dagger}|0,\widetilde{0}\rangle$ for bosons and $|1\rangle\rangle := |0,\widetilde{0}\rangle + ia^\dagger\tilde{a}^\dagger|0,\widetilde{0}\rangle$ for fermions. By using the thermal state $|\Psi_\alpha(t)\rangle\rangle$, we can obtain the mean value (*i.e.* finite temperature expectation value) of the microscopic description of physical quantity (denoted by A) at time t:

$$\langle A\rangle_t = \mathrm{Tr}\{\rho(t)A\} = \langle\langle\Psi_\alpha(t)|A|\Psi_\alpha(t)\rangle\rangle, \tag{2}$$

where $\langle\langle\Psi_\alpha(t)|$ is defined by $\langle\langle\Psi_\alpha(t)| := \langle\langle 1|\rho^{1-\alpha}$. Here and in the following, time dependence of the expectation values $\langle\cdots\rangle_t$ is indicated by a subscript t. Entropy $S(t)$ may be expressed by the density operator $\rho(t)$:

$$S(t) = -k\mathrm{Tr}\{\rho(t)\ln\rho(t)\} = \mathrm{Tr}\{\rho(t)[-k\ln\rho(t)]\} = \langle -k\ln\rho(t)\rangle_t, \tag{3}$$

where k is the Boltzmann constant. This equation defines entropy for nonequilibrium processes.

In what follows, it will be convenient to consider a dynamical variable

$$\eta(t) = -k\ln\rho(t) \tag{4}$$

as an *entropy operator* since its mean value $\langle\eta(t)\rangle_t$ coincides with the nonequilibrium thermodynamic (von Neumann) entropy $S(t)$ given by Eq. (3). Noticing that the density operator $\rho(t)$ can be written in terms of the entropy operator $\eta(t)$ as

$$\rho(t) = e^{-\eta(t)/k}, \tag{5}$$

the thermal state $|\Psi(t)\rangle\rangle$ in Eq. (1) can be expressed in terms of the entropy operator as

$$|\Psi_\alpha(t)\rangle\rangle = e^{-\alpha\eta(t)/k}|1\rangle\rangle. \tag{6}$$

The time derivative of Eq. (6) is given by

$$\frac{\partial}{\partial t}|\Psi_\alpha(t)\rangle\rangle = -\frac{\alpha}{k}\int_0^1 K(\lambda)\mathrm{d}\lambda|\Psi_\alpha(t)\rangle\rangle, \tag{7}$$

where

$$K(\lambda) = e^{-\lambda\eta}\frac{\partial\eta}{\partial t}e^{\lambda\eta} = \frac{\partial\eta}{\partial t} + \lambda\Big[\frac{\eta}{k}, \frac{\partial\eta}{\partial t}\Big] + \frac{\lambda^2}{2!}\Big[\frac{\eta}{k}, \Big[\frac{\eta}{k}, \frac{\partial\eta}{\partial t}\Big]\Big] + O(\lambda^3). \tag{8}$$

Meanwhile, in TFD the thermal state evolves with time according to the Schrödinger-like equation [10, 11]:

$$\frac{\partial}{\partial t}|\Psi_\alpha(t)\rangle\rangle = -\frac{i}{\hbar}\widehat{\mathcal{H}}(t)|\Psi_\alpha(t)\rangle\rangle, \tag{9}$$

where $\widehat{\mathcal{H}}(t) := \widehat{H} + i\widehat{\Gamma}(t)$ ($\widehat{H} := H - \widetilde{H}$) denotes a total-Hamiltoinan and $\hbar$ is Planck's constant. The operator H defined on $\mathscr{H}$ is a hermitian operator denoting the Hamiltonian of a system, and Γ defined on $\mathscr{L}$ is an arbitrary operator satisfying $(\widehat{\Gamma})^\sim = \widehat{\Gamma}$.

A general relation that connects the entropy operator $\eta(t)$ (defined by Eq. (4)) and the time dependent total Hamiltonian $\widehat{\mathcal{H}}(t)$ can be obtained from Eqs. (7) and (9) as

$$\int_0^1 K(\lambda)\mathrm{d}\lambda = \frac{ik}{\alpha\hbar}\widehat{\mathcal{H}}(t). \tag{10}$$

This is the most general EHR. It should be noted however that the time-dependent total Hamiltonian $\widehat{\mathcal{H}}(t)$ depends on the particular physical situation under study. Taking the expectation value of Eq. (10) and noticing $\langle H - \widetilde{H}\rangle_t = 0$, the EHR in Eq. (10) can be expressed as

$$\int_0^1 \langle K(\lambda)\rangle_t\,\mathrm{d}\lambda = -\frac{k}{\alpha\hbar}\langle\widehat{\Gamma}(t)\rangle_t. \tag{11}$$

Accordingly, if we assume $[\eta(t), \partial\eta(t)/\partial t] = 0$, the EHR in Eq. (11) can be expressed in a simple form:

$$\Big\langle\frac{\partial\eta(t)}{\partial t}\Big\rangle_t = -\frac{k}{\alpha\hbar}\langle\widehat{\Gamma}(t)\rangle_t. \tag{12}$$

It should be noted here that owing to the $\langle H - \widetilde{H}\rangle = 0$ setting in TFD, the EHR is expressed in terms of the entropy operator $\eta(t)$ and an arbitrary operator $\widehat{\Gamma}(t)$ only. In other words, we can say that the entropy production can be evaluated from the microscopic description of the operator $\widehat{\Gamma}(t)$. These general formulas expressing the EHR can be applied for open quantum systems.

Table 1. Mapping.

Open Quantum Systems	Thermo Field Dynamics
system: H_{sys}	nontilde: H
environment: H_{environ}	tilde: $\widetilde{H}$
interactions: $H_{\text{sys-environ}}$ (system-environment)	interactions: $H_{\text{nontilde-tilde}}$ (nontilde-tilde)

Let us consider an open quantum system within the framework of TFD in order to utilize the formulas obtained above. It is then worth reconsidering the physical foundation of TFD. We consider an open quantum system described by the Hamiltonian:

$$H_{\text{tot}}(t) = H_{\text{sys}}(t) + H_{\text{environ}}(t) + H_{\text{sys-environ}}(t), \tag{13}$$

where we assume $[H_{\text{sys}}, H_{\text{environ}}] = 0$ (*i.e.* the system and its environment are uncorrelated in the absence of the interactions between them). When this open system is in a thermal equilibrium state, the system is thermally equilibrium with its environment at the same temperature. In such a case, it is by no means to discriminate the state of the system thermodynamically from that of its environment, meaning that the environment in equilibrium is of de-individuation from the system in equilibrium [8]. Therefore, we do not care about the specific form of the environment Hamiltonian as far as those systems (the system of interest and its environment) are in the same thermal equilibrium state. Accordingly, the environment having the same thermodynamic property of the system can be regarded as the copy of the system of interest. Indeed, this allows us to assign the system of interest to the non-tilde space $\mathscr{H}$ and its environment to the tilde space $\widetilde{\mathscr{H}}$ in the representation space $\mathscr{L} := \mathscr{H} \otimes \widetilde{\mathscr{H}}$ introduced in TFD. Regarding the environment as the copy of the system, we can make the transformation of the system under consideration onto the non-tilde Hilbert space $\mathscr{H}$ (*i.e.* $H_{\text{sys}} \mapsto H$) and the environment onto the tilde Hilbert space $\widetilde{\mathscr{H}}$ (*i.e.* $H_{\text{environ}} \mapsto \widetilde{H}$). With this prescription, a natural transformation of an open quantum system within the theoretical structure of TFD can be accomplished by the following way: the interaction Hamiltonian $H_{\text{sys-environ}}$ is mapped to the representation space $\mathscr{L}$ by the mapping $H_{\text{sys-environ}} \mapsto H_{\text{nontilde-tilde}}(\equiv i\widehat{\Gamma})$. The mapping of an open system to TFD is summarized in Table 1.

In the following, we assume the case where the system is very close to the equilibrium state of its environment (in other words, the interaction between the system and its environment is very weak). In such a case we can safely employ the mapping described above for the system and its environment and we generalize Eqs. (11) and (12) for the case where there exists some interactions between the system and its environments. For an open quantum system, the interaction between the system and its environment possibly plays a special role in the dynamics (*e.g.* the spontaneous, irreversible thermodynamic processes) of the system. Obviously if there is some interaction between them, the system may be in non-equilibrium states. Whether the system is equilibrium or non-equilibrium due to the interaction can be judged from the study of entropy $S(t)$. Hereafter we assume the entropy for an open quantum system is evaluated from

$$\left\langle \frac{\partial \eta(t)}{\partial t} \right\rangle_t \equiv \frac{\partial \langle \eta(t) \rangle_t}{\partial t} = \frac{\partial S(t)}{\partial t}, \tag{14}$$

where $\langle \partial \eta(t)/\partial t \rangle$ is given by Eq. (12).

By applying the prescription for the transformation given above, Eqs. (11) and (12) are able to express an entropy production rate induced by *e.g.* the spontaneous or irreversible process, and therefore relate the entropy of the system to a possible interaction described by the microscopic interaction Hamiltonian $\widehat{\Gamma}$. The general formulas (11) and (12) will be used to investigate the effect of interaction directly on the system by the spontaneous or irreversible process for open quantum systems by giving the explicit form of interactions expressed by non-tilde and tilde operators. It should be noted that this interaction Hamiltonian $H_{\text{sys-environ}}(t)$ can be represented by the transformation $H_{\text{sys-environ}} \mapsto H_{\text{nontilde-tilde}}$ in the representation space $\mathscr{L}$ of TFD as expressed by the operators comprised of non-tilde and tilde operators.

3. Distribution function

Entropy is often expressed in terms of a distribution function $f(t)$:[1]

$$S(t) = -k\{f(t)\ln f(t) \mp [1 \pm f(t)]\ln[1 \pm f(t)]\}, \tag{15}$$

where the upper sign is for bosons while the lower sign is for fermions.

In this section, we will derive the expression for the entropy production in terms of a distribution function and also derive an equation of motion for a distribution function for open quantum systems.

Let us consider the entropy given by Eq. (15). Differentiating the entropy with respect to time t, we immediately obtain the expression for the entropy production of the system in terms of the distribution function $f(t)$:

$$\frac{\partial S(t)}{\partial t} = \begin{cases} -k\dfrac{\partial f(t)}{\partial t}\ln\dfrac{f(t)}{1+f(t)} & \text{(bosons)}, \\[2ex] -k\dfrac{\partial f(t)}{\partial t}\ln\{f(t)[1-f(t)]\} & \text{(fermions)}. \end{cases} \tag{16}$$

This indicates that the entropy for an open system can be determined by the knowledge of the distribution function only. Equating Eq. (12) to Eq. (16) and noting Eq. (14), we also obtain the equation of motion for the distribution function $f(t)$:

$$\hbar\frac{\partial f(t)}{\partial t} = \begin{cases} \dfrac{\langle\widehat{\Gamma}(t)\rangle_t}{\alpha\ln\frac{f(t)}{1+f(t)}} & \text{(bosons)}, \\[3ex] \dfrac{\langle\widehat{\Gamma}(t)\rangle_t}{\alpha\ln\{f(t)[1-f(t)]\}} & \text{(fermions)}. \end{cases} \tag{17}$$

It should be noted that the distribution function $f(t)$ for open systems could be determined with a given value of $\langle\widehat{\Gamma}(t)\rangle$ by solving the self-consistent Eq. (17) for $f(t)$.

4. Applications of EHR

In this section, we apply the EHR formula obtained in Sec. 2 to various types of boson model and consider how entropy is affected by the interactions between the system and its environment in open quantum systems. For the sake of convenience of calculations, we adopt the c-number parameter taken as $\alpha = 1$. It is noted that in the case $\alpha = 1$ the tilde operator and the non-tilde operator are related by the thermal state condition for the bra vacuum:

$$\langle\!\langle 1|a^\dagger = \langle\!\langle 1|\widetilde{a}, \tag{18}$$

where $\langle\!\langle 1| = \langle\!\langle \Psi_{\alpha=1}|$.

[1] The Bose or Fermi distribution function can be derived from Eq. (15) on the condition of the entropy maximum [12].

4.1. Dissipative interactions

Let us consider the interaction which causes energy dissipation and instability of the system [13, 14, 15]: In TFD, such an interaction may be expressed in the form of

$$\widehat{\Gamma} = g(t)(a^\dagger \tilde{a}^\dagger - \tilde{a}a). \tag{19}$$

We note that the above interaction ensures the condition $(\widehat{\Gamma})^\sim = \widehat{\Gamma}$ for the interaction Hamiltonians in open quantum systems. By substituting Eq. (19) into Eq. (12) and Eq. (17) and noticing Eq. (14), we obtain the equation for the entropy production rate:

$$\frac{\partial S(t)}{\partial t} = \left\langle \frac{\partial \eta(t)}{\partial t} \right\rangle_t = -\frac{k}{\hbar} g(t), \tag{20}$$

and the equation of motion for the distribution function:

$$\hbar \frac{\partial f(t)}{\partial t} = \frac{g(t)}{\ln \frac{f(t)}{1+f(t)}}, \tag{21}$$

respectively. From Eq. (20), the entropy at time t is expeditiously evaluated by

$$S(t) = S(0) - \frac{k}{\hbar} \int_0^t g(t) \mathrm{d}t. \tag{22}$$

It should be noted again that the entropy and the distribution function can be evaluated when the coupling function $g(t)$ of the system-environment interaction Hamiltonian [see Eq. (19)] is specifically given.

4.2. A relaxation process

A relaxation process is often described by the equation of motion for the distribution function [16]:

$$\hbar \frac{\partial f(t)}{\partial t} = -2\kappa \{f(t) - \bar{f}\}, \tag{23}$$

where κ is an arbitrary positive c-number parameter and $\bar{f}$ is an equilibrium distribution function specified by *e.g.* the Planck distribution function with temperature T:

$$\bar{f} = \frac{1}{e^{\hbar\omega/kT} - 1}. \tag{24}$$

Solving Eq. (23) for $f(t)$, we obtain

$$f(t) = \bar{f} - (\bar{f} - f_0) e^{-2\kappa t/\hbar}, \tag{25}$$

where f_0 is an initial distribution function.

Now we consider the Hamiltonian for an open quantum system, whose Hamiltonian in TFD can be expressed in a general form:

$$\widehat{\mathcal{H}} = H - \widetilde{H} + i\widehat{\Gamma}, \tag{26}$$

where we consider the case $H - \widetilde{H}$ and $\widehat{\Gamma}$ are specifically given by

$$H - \widetilde{H} = \hbar\omega(a^\dagger a - \tilde{a}^\dagger \tilde{a}), \tag{27}$$

$$\widehat{\Gamma} = -\kappa(a^\dagger a + \tilde{a}^\dagger \tilde{a} - 2\bar{f}) \left\{ \ln f(t) - \ln\big[1 + f(t)\big] \right\}. \tag{28}$$

Here we note that the $\widehat{\Gamma}$ in Eq. (28) ensures the condition $(\widehat{\Gamma})^{\sim} = \widehat{\Gamma}$. We can easily confirm that this Hamiltonian leads to the equation of motion for the distribution function given by Eq. (23) with the use of Eq. (17). Solving the self-consistent equations, Eqs. (23) and (28) for $f(t)$, we can obtain $f(t)$ and $\widehat{\Gamma}(t)$ at time t. Taking the same path to obtain Eq. (20), we obtain the equation for the entropy production rate for an open system:

$$\frac{\partial S(t)}{\partial t} = -\frac{k}{\hbar}\langle\widehat{\Gamma}(t)\rangle_t = \frac{2k\kappa}{\hbar}\{f(t) - \bar{f}\}, \tag{29}$$

where we used Eq. (28) for the interaction Hamiltonian. Substituting Eq. (25) into the right-hand side of Eq. (29), Eq. (29) can be written in the form:

$$\frac{\partial S(t)}{\partial t} = -\frac{2k\kappa}{\hbar}(\bar{f} - f_0)e^{-2\kappa t/\hbar}. \tag{30}$$

Furthermore integrating Eq. (30) with respect to time t, we obtain the entropy of the system at time t for the open quantum system described by the specific Hamiltonian $\widehat{\mathcal{H}}$ given by Eqs. (27) and (28):

$$S(t) = S(\infty) + k(\bar{f} - f_0)e^{-2\kappa t/\hbar}. \tag{31}$$

This equation implies that the entropy of the system will relax exponentially to $S(\infty)$ with the relaxation time $\tau := \hbar/2\kappa$ due to the presence of the interaction given by Eq. (28).

5. Summary and concluding remarks

We developed a microscopic theory of entropy and derived a general formula which connects entropy to a system Hamiltonian. This formula, so-called the "entropy-Hamiltonian relation" (EHR), is applied for an open quantum system with the help of the Liouville-space formulation or thermo field dynamics (TFD). When an open system is in a thermal equilibrium state, the system is thermally equilibrium with its environment at the same temperature. On the basis of the zeroth law of thermodynamics, it is by no means to discriminate the state of the system thermodynamically from that of its environment, meaning that the environment in equilibrium is of de-individuation from the system in equilibrium. As shown in Sec. 2, this allows us to relate the interactions to the entropy of the system and to investigate the effect of interactions in nonequilibrium open quantum systems by assuming that we can consider the environment to be a copy of the system. Therefore, we can assign the system of interest to the non-tilde space $\mathscr{H}$ and its environment to the tilde space $\widetilde{\mathscr{H}}$ in the representation space $\mathscr{L} := \mathscr{H} \otimes \widetilde{\mathscr{H}}$ introduced in TFD. Regarding the environment as the copy of the system, we can indeed make the transformation of the system under consideration onto the non-tilde Hilbert space $\mathscr{H}$ (*i.e.* $H_{\mathrm{sys}} \mapsto H$) and the environment onto the tilde Hilbert space $\widetilde{\mathscr{H}}$ (*i.e.* $H_{\mathrm{environ}} \mapsto \widetilde{H}$). With this prescription, we accomplished a natural transformation of an open quantum system within the theoretical structure of TFD in the following way: the interaction Hamiltonian $H_{\mathrm{sys-environ}}$ is mapped to the representation space $\mathscr{L}$ by the mapping $H_{\mathrm{sys-environ}} \mapsto H_{\mathrm{nontilde-tilde}}(\equiv i\widehat{\Gamma})$.

We derived the EHR formulas (11) and (12) which connect the interaction Hamiltonian to the entropy of the system in nonequilibrium open quantum systems. The equation of motion for the distribution function (17) is also obtained from the entropy represented by the distribution function. We apply the EHR formula and the equation of motion for a distribution function obtained in Sec. 2 to two different kinds of boson model. In particular, we considered the case where the interaction causes dissipation and the case where the interaction causes a relaxation process to show the applicability of obtained formulas. We obtained the entropy production by using these interaction Hamiltonians to the EHR formula. If we find the microscopic description of the interaction Hamiltonian $\widehat{\Gamma}$, all we need to do is to utilize the EHR formulas (11) and (12), and we will have the entropy production from which we can obtain some information about the interaction through the entropy of the system for the open quantum system.

Finally, one important remark is relevant here. The quantum effects play a very essential role in the nanoscale tribology [17, 18, 19, 20, 21]. For instance, the Caldeira-Leggett model introduced to elucidate the role of (quantal) friction is successfully formulated by considering a macroscopic quantum tunneling in terms of path-integral method [17]. However, their formulation is difficult to treat and calculate analytically. While, our theory is based on canonical formulation and therefore it can be more intuitively understood the interacting phenomena, and easier to treat and evaluate than the path-integral method. Indeed, our new formulation presented in this article (i.e. the EHR) enables us to calculate the entropy production directly from system-environment interactions. This means that it allows us to conjecture what type of interactions play an essential role for frictional effects on the system just using a conjectured form of the interaction Hamiltonian $\widehat{\Gamma}$.

References

[1] Breuer H P and Petruccione F 2002 *The Theory of Open Quantum Systems* (Oxford: Oxford University Press)
[2] Reents G 2002 *AIP Conf. Proc.* **643**, 25
[3] Takahashi Y and Umezawa H 1975 *Collect. Phenom.* **2** 55
[4] Umezawa H, Matsumoto H and Tachiki M 1982 *Thermo Field Dynamics and Condensed States* (Amsterdam: North-Holland)
[5] Kosakowski A, Frigerio A, Gorini V and Verri M 1977 *Commun. Math. Phys.* **57** 97
[6] Ojima I 1986 *Lett. Math. Phys.* **11** 73
[7] Ojima I 1989 *J. Stat. Phys.* **56** 203
[8] Majima H and Suzuki A 2005 *J. Kor. Phys. Soc.* **46** 678
[9] Majima H 2008 *AIP Conf. Proc.* **985** 756
[10] Suzuki M 1985 *J. Phys. Soc. Jpn.* **54** 4483
[11] Arimitsu T and Umezawa H 1985 *Prog. Theor. Phys.* **74** 429
[12] Landau L D and Lifshitz E M 1980 *Statistical Physics, Part 1. Vol. 5 (3rd ed.)* (Oxford: Butterworth-Heinemann)
[13] Feshbach H and Tikochinsky Y 1977 *Transactions of the New York Academy of Sciences* **38** 44
[14] Celeghini E, Rasetti M and Vitiello G 1992 *Ann. Phys.* **215** 156
[15] Majima H and Suzuki A *in preparation*
[16] Pitaevskii L P and Lifshitz E M 1981 *Physical Kinetics. Vol. 10 (1rst ed.)* (Oxford: Pergamon Press)
[17] Caldeira A O and Leggett A 1983 *Ann. Phys.* **149** 374
[18] Pal S 1989 *Phys. Rev. A* **39** 3825
[19] Hu B, Li B and Zhang W 1998 *Phys. Rev. E* **58** R4068
[20] Krajewski F R and Müser M H 2004 *Phys. Rev. Lett.* **92** 030601
[21] Inui N and Miura K 2010 *e-J. Surf. Sci. Nanotech.* **8** 57

A New Nanofiller to Improve the Friction and Wear Properties of PPS

Ryoichi Hiroi (1), Hitoshi Tanaka (2)

(1) Graduate School of Advanced Technology and Science, University of Tokushima, Minamijosanjima-cho, Tokushima 770-8506, Japan.

(2) Institute of Technology and Science, University of Tokushima, Minamijosanjima-cho, Tokushima 770-8506, Japan.

ryoteen@hotmail.co.jp

Abstract. An organically modified layered titanate (OMLT) was used as a new filler to prepare PPS-based nanocomposite (PPSNC). The melt mixing of PPS with OMLTs led to an exfoliation of stacked-layers and achieved the nanoscale dispersion of titanate layers in a PPS matrix. Wear resistance of the PPSNC was excellent as compared with the pristine PPS.

PPS/maleic anhydride polyethylene (MA-PE) alloy with OMLT were prepared to enhance the tribological property of PPS. OMLT successfully behaved as the inorganic compatibilizer for PPS/MA-PE alloy, and the polymer alloy had an excellent friction and wear properties.

1. Introduction

Poly(p-phenylensulfide) (PPS) is an excellent crystalline thermoplastic with flame and chemical resistance, and it is used broadly as an alternative material of metal in car parts, electricity and electronic products, and precision parts. To improve the tribological property is important for the durability of the materials. In a previous study, polymer alloy techniques were reported to improve the friction and wear properties [1].

Over the last few years, nanocomposite technology consisting of a polymer and organically modified layered silicate (OMLS) has developed. It has been reported that a nanocomposite has excellent wear resistance [2]. On the other hand, there has been a growing interest in applying nanocomposite technology in a polymer blend. Some researchers reported that OMLS could play a role as a compatibilizer for immiscible polymer blends [3].

2. Experimental results and Discussion

The $K_{0.7}Ti_{1.73}Li_{0.27}O_{3.95}$, which is layered titanate with a lepidocrosite-like structure, was modified with cocomethylbis(hydroxyethyl) ammonium salt. The obtained organically modified layered titanate (OMLT) was melt-mixed with PPS resin by twin-screw extruder, and the composite pellets were molded with an injection machine to prepare an ASTM test piece. Synthetic fluoromica modified by cocomethylbis(hydroxyethyl) ammonium (OMLS) was also melt-mixed with neat PPS for comparing. Hereafter, OMLT and OMLS filled composites are abbreviated as PPSNC1 and PPSNC2, respectively. Moreover, the PPSNC1 pellet was also melt-mixed with a maleic anhydride polyethylene (MA-PE) to prepare the PPS/MA-PE alloy with OMLT.

A high-resolution transmission electron microscopy (TEM) confirmed that the titanate layers were exfoliated uniformly with the thickness of nanoscale size in PPSNC1. The morphology of MA-EBR domain in PPS/MA-EBR alloy with OMLT was observed with a scanning electron microscopy (SEM).

Tribological tests were performed on a Suzuki-type friction machine to investigate the friction and wear properties.

2.1. Preparation of PPS-based nanocomposite (PPSNC)

TEM images of PPSNCs are shown in Figure 1. As seen in Figure 1(a), PPSNC1 with OMLT shows a fine and uniform nanoscale dispersion of titanate layers in the PPS matrix where the titanate particles in both exhibit perpendicular alignment to the sample surface. While, some stacked-silicate layers were observed in PPSNC2 with OMLS [Figure 1(b)]. The PPS-based nanocomposite was obtained only by using OMLT as a nano-filler in this condition.

Table 1 shows the tribological properties of PPSNC1. The PPSNC1 with 5%OMLT shows less wear rate compared with neat PPS, whereas the friction coefficient of PPS did not change at all.

Figure 1 TEM images of PPS-based composites; (a) PPSNC1 (b) PPSNC2

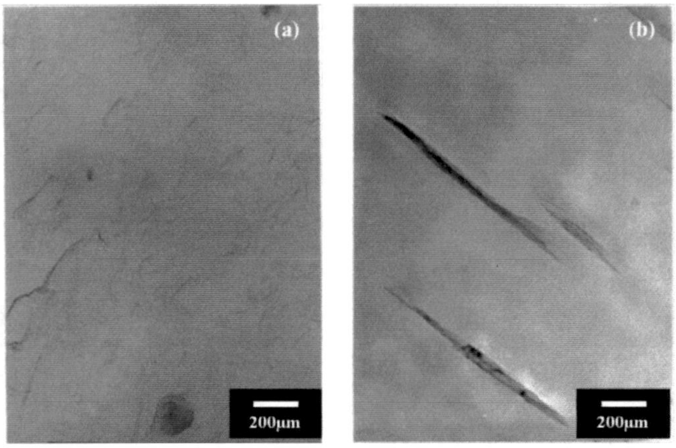

Table 1 Tribology of PPS-based nanocomposite with OMLT

	Friction coefficient (μk)	Wear rate (mm^3/m)
PPS	0.54	1.17E-05
PPSNC1	0.58	6.04E-06

2.2. Polymer alloy with the nanocomposite technique

SEM images of PPS/MA-PE and PPSCN1/MA-PE are shown in Figure 2. These SEM images show typical sea-island morphologies. In the sea-island structure, the continuous phase and the domains indicate PPS (or PPSNC) and MA-PE, respectively. The average domain size and the interfacial distance estimated from those SEM images are summarized in Table 2. PPS/MA-PE alloy exhibits poor dispersion of MA-PE domains, with a large domain size, and large distance between domains [Figure 2(a), and Table 2]. In contrast to the above, the decreasing of the domain size and the interfacial distance between each domain were observed in PPSNC/MA-EBR [Figure 2(b), and Table 1]. From the result, it is obvious that a use of PPSNC as a matrix polymer governs the morphologies of domains in PPS-based polymer blends. Consequently, it means that OMLT behaved as an effective compatibilizer for a polymer alloy, although the titanate layers are an inorganic material.

Figure 2 SEM images of PPS/MA-PE and PPSNC1/MA-PE:
(a) PPS/MA-PE (80/20) (b) PPSNC1/MA-PE (80/20)

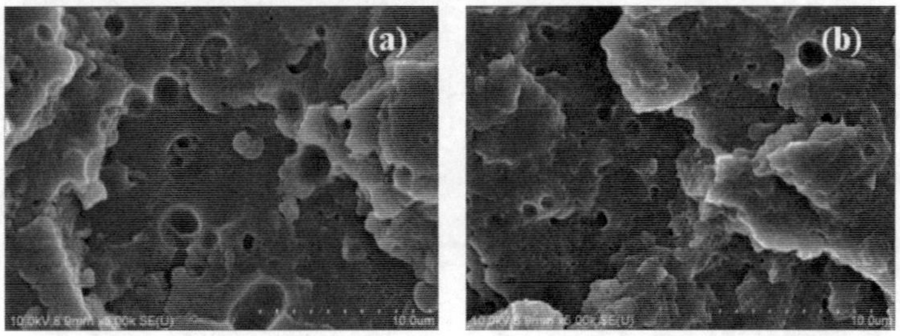

Table 2 Average domain size and interfacial distance observed from SEM images

Composition wt%/wt%	Domain size / μm	Interfacial distance/ μm
PPS/MA-PE 80/20	1.4	1.3
PPSNC1/MA-PE 80/20	0.6	0.6

The tribological property of PPSNC1/MA-PE alloy is shown in Table 3. Moreover, Figure 3 shows the friction coefficient versus sliding distance at a constant applied load of 20kgf/mm^2 for the PPS/MA-PE and PPSNC1/MA-PE alloys. The tribological properties improved generally by blending the MA-PE. In addition, the PPSNC1/MA-PE, in which PPSNC1 was used as a matrix polymer, also showed less wear rate compared with PPS/MA-PE (Table 3). It was indicated that the sliding of PPSNC1/MA-PE compared with the PPS/MA-PE was stabilized obviously as seen in Figure 3, although the mean friction coefficient was hardly different between PPS/MA-PE and PPSNC1/MA-PE in Table 3. Presumably, the stabilized sliding property has been brought by the decreasing of the wear rate.

Table 3 Tribology of PPS/MA-PE alloy with OMLT

	Friction coefficient (μk)	Wear rate (mm^3/m)
PPS/MA-PE	0.29	3.56E-07
PPSNC1/MA-PE	0.27	2.96E-08

Figure 3 Tribological test of PPS/MA-PE and PPSNC1/MA-PE: (a) PPS/MA-PE (b) PPSNC1/MA-PE

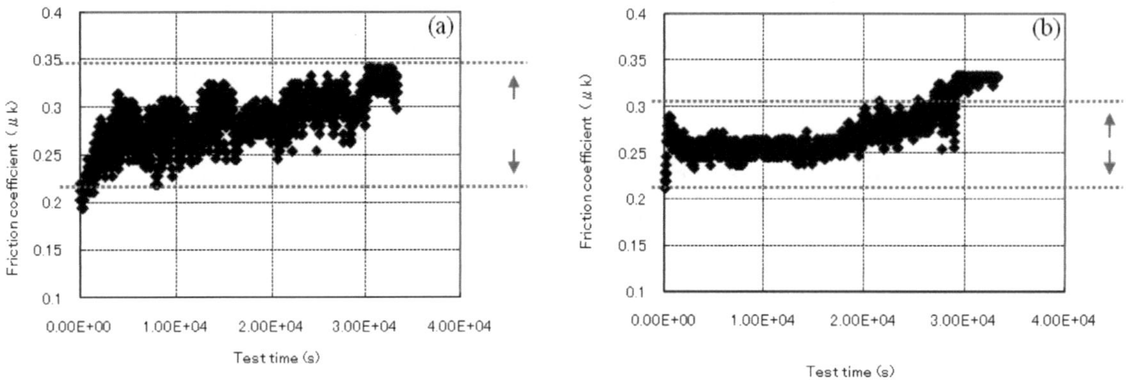

3. Conclusion

The friction and wear properties were improved by using a polymer alloy with a nanocomposite technique. Namely, the OMLT behaved as a compatibilizer to decrease a domain size of MA-PE in PPS/MA-PE, while improving the wear property. We conclude that the OMLT is an effective filler to improve the tribological properties of the PPS, because the OMLT can be successfully exfoliated in the PPS matrix.

References

[1] Horiuchi S and Ishii Y 2000 *Polymer Journal* **32** 339
[2] Mu B, Wang O, Wang T, Wang H and Jian L 2008 *Polymer Eng. and Sci.* **48** 203
[3] Wang Y, Zhang Q and Fu Q 2003 *Macromol. Rapid Commun.* 24 231

International Conference on Science of Friction 2010 (ICSF2010) IOP Publishing
Journal of Physics: Conference Series **258** (2010) 012017 doi:10.1088/1742-6596/258/1/012017

Estimation of Interfacial and Adhesion Energies of Anthracene/Graphite (0001) system Based on Wulff's Theorem

S Jo[1] and M Takenaga[2]
[1]Division of General Education, Aichi Gakuin Univeristy,
12 Araike, Iwasaki-cho, Nisshin, Aichi 470-0195, Japan
[2]Department of Applied Chemistry, Tokyo University of Science, Yamaguchi,
1-1-1 Daigaku-Dori, Sanyoonoda, Yamaguchi 756-0884, Japan

[1]sadaharu@dpc.agu.ac.jp

Abstract. The application of Wulff's theorem to an anthracene/graphite (0001) system is discussed in this paper. Anthracene crystals on a graphite (0001) substrate was prepared by using physical vapor growth technique with a horizontally arranged furnace. Many crystals showed line coincidences at the interface between crystals and graphite substrate; $[010]_{anthracene}//[\bar{2}110]$, $[1\bar{2}10]$, or $[11\bar{2}0]_{graphite}$, and $(100)_{anthracene}//(0001)_{graphite}$, and a few maintained $[010]_{anthracene}//[\bar{2}110]$, $[1\bar{2}10]$, or $[11\bar{2}0]_{graphite}$, and $(001)_{anthracene}//(0001)_{graphite}$. Most crystals grew as the typical Wulff's polygons on a graphite substrate and they kept the $(100)_{anthracene}/(0001)_{graphite}$ relationship with cutting off near half of the $(001)_{anthracene}$ plane. Such morphology enables the energetic estimation based on the Wulff's theorem. Resultantly, both interfacial and adhesion energies of the $(100)_{anthracene}/(0001)_{graphite}$ system could be calculated to be 63.1 ~ 65.7 and 50.0 ~ 52.6 erg cm^{-2}, respectively.

1. Introduction

Plastic devices have a lot of possibilities for applying to flexible electronics of next-generation. The single crystals grown via the physical vapor growth technique have been much focused on various types of device applications in recent years. Karl firstly obtained 10-mm-sized organic single crystals using physical vapor growth technique with carrier gases [1]. After that, a horizontal arranged growth furnace had been developed by Kloc et al. [2] and Laudise et al. [3], and they succeeded in the growth of several kinds of organic single crystals, for example thiophene-oligomers, pentacene, tetracene, anthracene. Using physical vapor growth technique with horizontally arranged furnace, Jo et al. firstly succeeded in the construction of the line coincidence between anthracene crystals and a graphite (0001) substrate [4]. It is usual that the ultra thin organic films with line coincidencies can be obtained on a foreign substrate with using a high-vacuum system, but the physical vapor growth technique is also applicable for the same aim. Especially, in case of the anthracene/graphite system, quite fine relation for line coincidence can be achieved via the physical vapor growth. Moreover, it have been cleared that many crystals kept the $(100)_{anthracene}/(0001)_{graphite}$ relation with cutting off near half of the $(001)_{anthracene}$ plane, when they were obtained by physical vapor growth technique. Such morphology

[1] Corresponding Author: sadaharu@dpc.agu.ac.jp (S. Jo)

seems to be suitable to estimate several crystallographic energies regarding their interfaces based on the Wulff's theorem. The equilibrium situation corresponds to one crystal with {hkl} planes exposed such that the total surface free energy for the given volume was lead as the minimization, where the surface free energies $\sigma(hkl)$ depend on the crystal orientation. Above statement is well known as the Wulff's theorem [5]. The equilibrium shape of a three-dimensional free crystal can be obtained as the envelop of the planes with the normal distance from the Wulff's point proportional to the difference of the corresponding specific surface free energy. The equilibrium shape of epitaxially orientated crystals will be formed on a substrate so as to minimize the total energy of the system. Under such condition, the Wulff's construct can be applied to the crystal-substrate system. In this paper, the application of Wulff's theorem to an anthracene/graphite (0001) system will be treated quantitatively.

2. Experimental

Anthracene crystals on a graphite (0001) substrate was prepared by using physical vapor growth technique with a horizontally arranged furnace as indicated in Figure 1. A graphite (0001) substrate was set at the growth area of the growth tube, and newly cleaved under helium gas atmosphere. The depositant anthracene was purchased from Tokyo Kasei Co. Ltd. A reaction tube, a crystal growth tube, a source boat, a resistance wire and a band heater were used for the construction of the growth system. Both lengths and diameters of the reaction tube and growth tube are 500 and 30 mm, and 250 and 25 mm, respectively. Small tubes with a diameter of 5 mm were set at both edges of the reaction tube for inflow and outflow of helium carrier gas. The evaporation temperature was set at 120 °C and the growth time was set at 500 hours to obtain crystals under the quasi-static condition. After the crystal growth, the substrate with many crystals was substantially removed from the furnace, and morphological and geometrical investigations were performed by using optical microscope and scanning probe microscope (SEM). Both interfacial and adhesion energies were estimated based on Wulff's theorem.

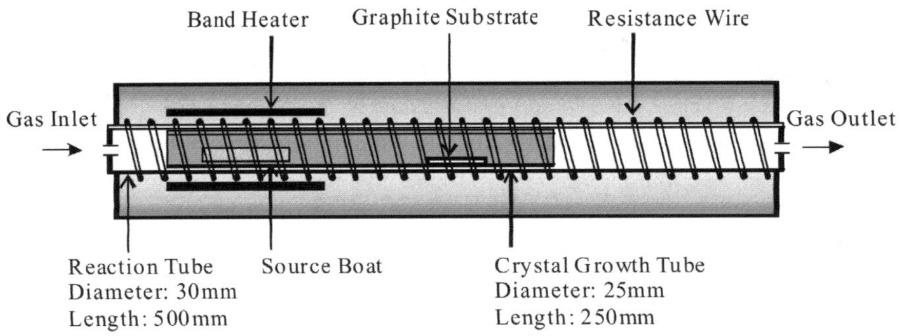

Figure 1. Horizontally arranged furnace system for physical vapor growth.

3. Results and Discussion

Figure 2 shows a photograph of an anthracene crystal without any substrate obtained by physical vapor growth technique. The crystals indicated a plateletlike morphology with a pair of large hexagonal planes, and their morphologies were identical irrespective of their sizes. Based on X-ray diffraction experiments, the large hexagonal plane could be determined as (001) [4]. The faceted angles in the basal (001) plane originate from the crystallographic angles between the lateral (110), $(\bar{1}$

10), (1$\bar{1}$0), ($\bar{1}\bar{1}$0), (100), and ($\bar{1}$00) planes, as indicated in Figure 3. The morphology well agrees with the theoretical result deduced from surface energy calculations performed by Northrup et al. [6].

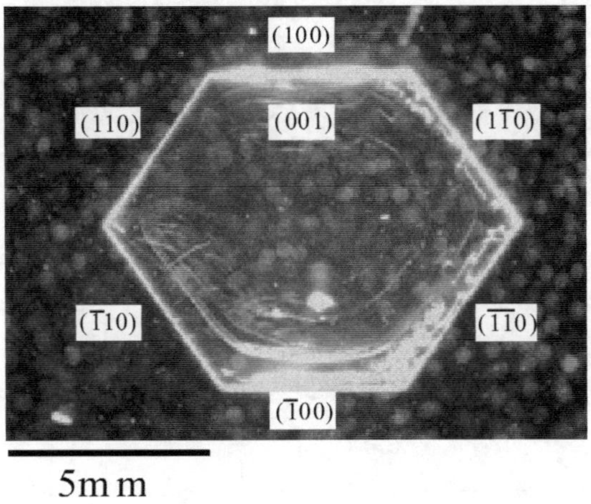

Figure 2. Photograph of an anthracene crystal obtained without any foreign substrate.

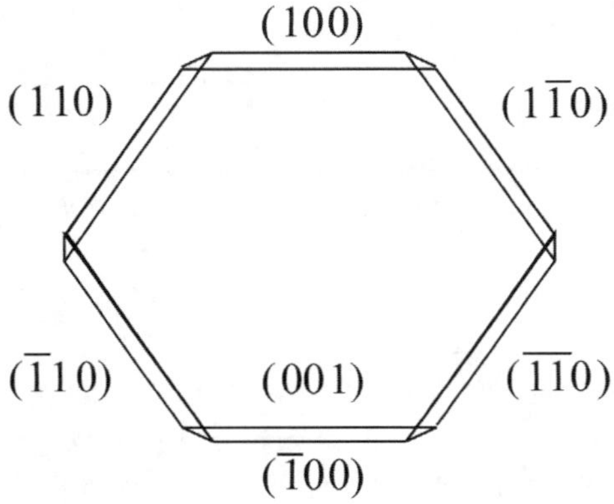

Figure 3. Configuration of anthracene crystal.

Figure 4 shows a photograph of anthracene crystals grown on a graphite (0001) substrate. Each edges of a large triangle drawn with white lines indicate the crystallographic directions of the graphite substrate: [2$\bar{1}\bar{1}$0], [$\bar{1}$2$\bar{1}$0], or [11$\bar{2}$0]$_{graphite}$. Many crystals maintained following line coincident relationships; [010]$_{anthracene}$//[2$\bar{1}\bar{1}$0], [$\bar{1}$2$\bar{1}$0], or [11$\bar{2}$0]$_{graphite}$, and (100)$_{anthracene}$//(0001)$_{graphite}$, and a few maintained [010]$_{anthracene}$//[2$\bar{1}\bar{1}$0], [$\bar{1}$2$\bar{1}$0], or [11$\bar{2}$0]$_{graphite}$, and (001)$_{anthracene}$//(0001)$_{graphite}$. Most crystals grew with keeping typical Wulff's polygons on a graphite substrate. Many roofs of the crystals seen in Figure 4 correspond to (100) planes. Figure 5 shows a microscopic image of an anthracene crystal grown on a (0001)$_{graphite}$ substrate observed by using scanning electron microscope (SEM). The crystal kept the (100)$_{anthracene}$/(0001)$_{graphite}$ relation with cutting off near half of the (001)$_{anthracene}$ plane and such morphology indicates the typical Wulff's polygon. Figure 6 shows a schematic drawing of an anthracene crystal on a graphite substrate. The Wulff' point is denoted by the symbol O. The distance from O to the upper (100) plane is denoted by h_1. The distance from the upper (100) plane to the contact plane and that from O to the contact plane are denoted by h_A and h_{OB}, respectively. Each symbols σ_A, σ_B, σ_{AB} and γ correspond to the surface energy of the (100)$_{anthracene}$, surface energy of the (0001)$_{graphite}$, interface energy and adhesion energy of this system, respectively. The hypothetic part of the crystal appearing in case of the absence of a foreign substrate is shown by the additional dashed lines.

Generally, γ governs the combination of planes between depositants and the foreign substrate; a plane of a depositant with the maximum γ lies on the substrate. The ratio h_{OB}/h_1 depends on the magnitudes of both the adhesion γ and interfacial σ_{AB} energies. The (100)$_{anthracene}$-(0001)$_{graphite}$ interfacial and adhesion energies can be calculated, when the ratio h_{OB}/h_1 is known accuracy, provided that the surface free energy σ_A of the (100)$_{anthracene}$ is known and also that σ_B of the

$(0001)_{graphite}$ plane. The most commonly assumed values for the surface free energies of $(100)_{anthracene}$ and $(0001)_{graphite}$ are 65.7 erg cm^{-2} 6) and 50 erg cm^{-2} [7,8], respectively. By taking 50 erg cm^{-2} as the surface free energy σ_B, the interfacial energy can be calculated thanks to the following equation, which can be used effectively when a crystal is formed on a foreign substrate:

$$\frac{\sigma_n}{h_n} = \frac{\sigma_A}{h_1} = \frac{\sigma_{AB} - \sigma_B}{h_{OB}} = \frac{\sigma_A - \gamma}{h_{OB}} = const. \quad (1)$$

,where (1) includes the well-known relation of Dupre:

$$\sigma_{AB} = \sigma_A + \sigma_B - \gamma \quad (2)$$

The relationship (1) follows that the distance h_n from the Wulff's point to the contact plane is proportional to the difference of the corresponding specific surface energy and specific adhesion energy. The equation (1) can be transformed as follows;

$$\frac{h_{OB}}{h_1} = \frac{\sigma_{AB} - \sigma_B}{\sigma_A} \quad (3)$$

Hence, determining the ratio h_{OB} / h_1 experimentally, we can obtain the value of σ_{AB} with using established σ_A and σ_B. For determining the ratio h_{OB} / h_1 accuracy, many SEM images of anthracene crystals on graphite (0001) substrate were used, and the morphologies of anthracene crystals were investigated in detail comparing Wulff's construction. Finally, the ratio h_{OB} / h_1 could be determined as 0~ -0.2 experimentally. Here, $h(100)<0$ because the Wulff's point O exists under or on the interface. Such values of the ratio h_{OB} / h_1 seems to be reasonable because each surface free energies of $(100)_{anthracene}$ 65.7 erg cm^{-2} and $(0001)_{graphite}$ 50.0 erg cm^{-2} are very close each other. Based on the formula (1), Interfacial energy σ_{AB} was calculated as σ_{AB}= 36.9~50.0 erg cm^{-2}. Also, the adhesion energy γ could be calculated by Dupre's equation (2) as following: $\gamma = \sigma_A + \sigma_B - \sigma_{AB}$ =65.7~78.8 erg cm^{-2}.

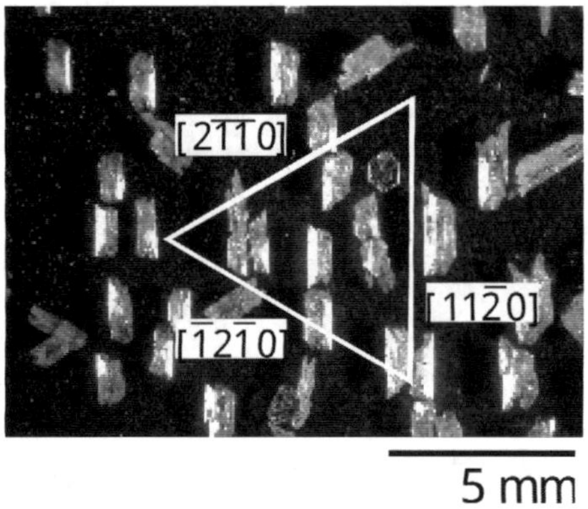

Figure 4. Linearly aligned anthracene crystals arranged on a $(0001)_{graphite}$ substrate.

In case of anthracene/graphite(0001) system, the large $(001)_{anthracene}$ plane never contact to the $(0001)_{graphite}$ substrate because the surface energy of (001)anthracene plane is relatively lower than that of the $(100)_{anthracene}$ plane, the surface energies of the $(001)_{anthracene}$ and $(100)_{anthracene}$ are 3.3 and 4.1 meV/Å^2, respectively [6]. Because the surface energy of the $(100)_{anthracene}$ plane is higher than that of the $(001)_{anthracene}$, the building-like morphologies of anthracene crystals on $(0001)_{graphite}$ substrate could be constructed by physical vapor transport technique. Moreover the line coincidency between the $(001)_{anthracene}$ and $(0001)_{graphite}$ plane help to construct such building-like morphologies. The $d_{(001)}$ is equal to 9.7 Å and the unit vectors of $(0001)_{graphite}$ plane is 2.1 Å. Near five times coincidency might appear for the interface between each planes of the $(001)_{anthracene}$ and $(0001)_{graphite}$.

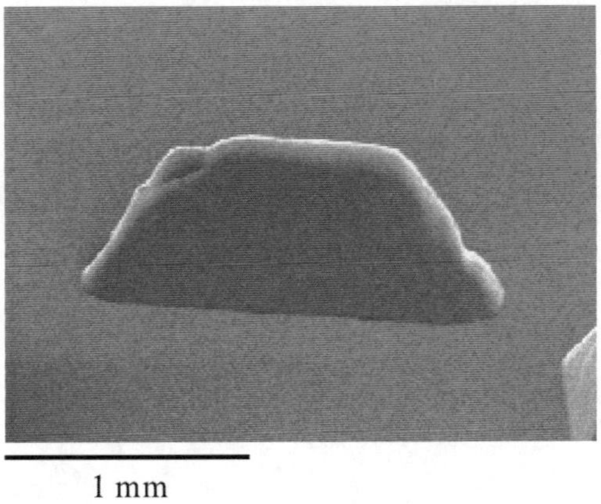

Figure 5. Anthracene crystal with the shape with cutting off near half of the $(001)_{anthracene}$ plane observed by SEM.

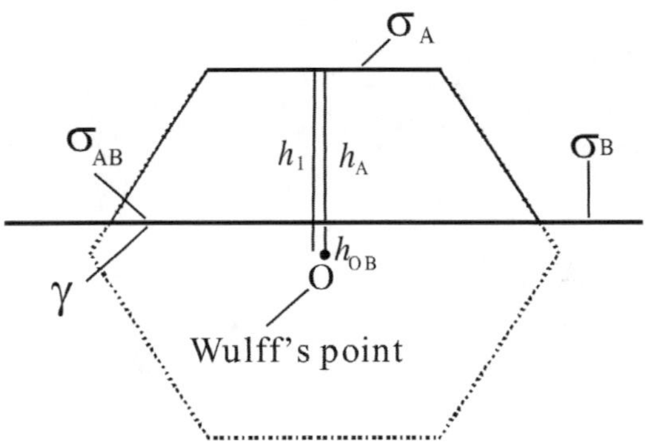

Figure 6. Schematic drawing of an anthracene crystal on a graphite (0001) substrate.

4. Conclusion

In conclusion, anthracene crystals with the equilibrium shape were fabricated on a $(0001)_{graphite}$ substrate by using physical vapor growth technique. Most anthracene crystals kept a manner with following epitaxial relation; $[010]_{anthracene}//[2\bar{1}\bar{1}0]$, $[\bar{1}2\bar{1}0]$, or $[11\bar{2}0]_{graphite}$, and the $(100)_{anthracene}/(0001)_{graphite}$ relation with cutting off near half of the $(001)_{anthracene}$ plane. Based on the Wulff's theorem with using established values surface free energies of $(100)_{anthracene}$ and $(0001)_{graphite}$, the $(100)_{anthracene}$-$(0001)_{graphite}$ interfacial and adhesion energies could be estimated as 36.9~50.0 and 65.7~78.8 erg cm^{-2}, respectively. This research was partly supported by the Grant-in-Aid for Young Scientists (B), No.19670012, The Ministry of Education, Culture, Sports, Science and Technology (MEXT), Japan.

The author wish to thank Prof. M. Yoshimura (Toyota Technological Institute) for his kind experimental support. SEM experiments were performed at Surface Science Laboratory, Toyota Technological Institute with the approval of Nanonet, The Ministry of Education, Culture, Sports, Sceience and Technology (MEXT), Japan.

References

[1] N. Karl: J. Cryst. Growth **187** (1990) 1009.
[2] Ch. Kloc, P. G. Simpkins, T. Siegrist, and R. A. Laudise: J. Cryst. Growth **182** (1997) 416.
[3] R. A. Laudise, Ch. Kloc, P. G. Simpkins, and T. Siegrist: J. Cryst. Growth **187** (1998) 449.

[4] S. Jo, H. Yoshikawa, A. Fujii, and M. Takenaga: Jpn. J. Appl. Phys. **44** (2005) 4187.
[5] G. Wulff: Z. Krist., **34** (1901) 449.
[6] J. E. Northrup, M. L. Tiago, and S. G. Louie: Phys. Rev. B **66** (2002) 121404.
[7] R. J. Good, L. A. Girifalco and G. Kraus: J. Phys. Chem. **62** (1958) 1418.
[8] B. Bilinski: Mater. Chem. Phys. **18** (1987) 231.

Mechanical response of Kr films adsorbed on graphite substrate with a quartz tuning fork

Hajime Kobayashi[1], Junko Taniguchi[1], Masaru Suzuki[1], Kouji Miura[2], and Ichiro Arakawa[3]

[1] Department of Engineering Science, University of Electro-Communications, Chofu, Tokyo 182-8585, Japan.
[2] Department of Physics, Aichi University of Education, Kariya, Aichi 448-8542, Japan.
[3] Department of Physics, Gakushuin University, Toshima, Tokyo 171-8588, Japan

E-mail: `hajime@phys.uec.ac.jp`

Abstract. We have developed a quartz-crystal microbalance (QCM) using a quartz tuning fork resonator, and have carried out experiments for Kr lms adsorbed on a single crystalline graphite at LN_2 temperature. For the monolayer lm, the coupled mass to the oscillating substrate is signi cantly small until the lm enters the commensurate (C) phase. It starts to increase at the C phase, and the increase in the coupled mass with respect to the coverage decreases at the coexistence phase between the C and the incommensurate (IC) phases. It is enhanced again when the lm enters the IC phase. We compared the observed behavior with that of 5 MHz QCM experiments, and found that the sliding motion does not depend on oscillation frequency.

1. Introduction

Sliding motion of physisorbed lms on the lateral oscillating substrate has attracted interest because of a study on the atomic-scale mechanism of sliding friction. Noble gas lms adsorbed on graphite take the commensurate (C) and the incommensurate (IC) phases by changing the fractional coverage. By using these lms, it is possible to make clear the relation between the sliding friction and the commensurability of lm.

We have already reported 5 MHz AT-cut quartz-crystal microbalance (QCM) experiments for Kr lms adsorbed on a single-crystalline graphite (Kr/Gr) [1]. It was found that the coupled mass starts to increase at the C phase, and that this increase continues after the commensurate-incommensurate (C-IC) transition. The observed behavior is di erent from a naive idea that lms of the C phase do not slide easily because of their structural pinning [2].

Thus motivated, we planned to measure the mechanical response of Kr/Gr when an inertial force acting on lm is small. Since the inertial force is proportional to the product of the oscillation amplitude and the square of the angular frequency in the QCM technique, we developed QCM using a quartz tuning fork resonator, and carried out experiments for Kr/Gr. In this paper, we explain the present results of a 32 kHz quartz tuning fork of Kr/Gr, focusing on two topics: one is the relation between the coupled mass and the coverage of lm, and the other is the sliding direction dependence. We also compare them with those of 5 MHz AT-cut QCM experiments.

2. Experimental

A quartz tuning fork oscillates exurally perpendicular to the arms. In QCM using the tuning fork, the substrate of lm is attached at the tips of the arms. When the lm adsorbed on the substrate moves in concert with the tuning fork, the resonance frequency decreases from that of no lms is expressed as

$$\frac{\Delta f}{f} \sim \; 2\frac{m}{M},$$

where m is the coupled mass of lm and M is the mass of the arms. In the present experiments, the resonator is a 32 kHz quartz tuning fork.

The experimental procedure is similar to 5 MHz AT-cut QCM experiments [1]; here, we describe it brie y. Single-crystalline graphite was rst obtained from a kind of mineral called Franklin Marble, by dissolving it out with hydrochloric acid. To exfoliate, a piece of graphite was immersed in the reaction mixture of concentrated sulfuric acid and nitric acid under stirring for 16 h. After the neutralization and dehydration, the interlayer space was expanded by heating in furnace at 1050°C for 15 s. For cleaning, the exfoliated graphite was heat-treated in a vacuum at 900°C for 4 h. The Laue photograph of the exfoliated graphite is shown in Fig. 1 (a), and demonstrates that it remains the crystal orientation in the a-b plane.

To prepare a quartz tuning fork with graphite, 200 nm-thick Ag lm was deposited on the graphite. Then, the tuning fork and Ag-plated graphite were pressed together and were heated in hydrogen atmosphere at 300°C for 1 h. After bonding, any excess graphite was carefully removed. The tuning fork with graphite is shown in Fig. 1 (b). The e ective surface area of graphite on the arms is about 2 cm^2, and the Q value of the tuning fork was higher than 2×10^4 at LN_2 temperature.

We prepared two quartz tuning forks with graphite: one of the tuning forks oscillates parallel to the a-axis of graphite, while the other in the direction inclined at about 30° with respect to the the a-axis. The tuning forks were set in a sample cell. In addition to the tuning forks, Grafoil disks (exfoliated graphite) were set in the cell in order to increase the surface area and to minimize the e ect of the desorption of Kr. After setting, the cell was sealed with helico ex (HN100), and was evacuated at 120 °C for 96 h to remove water, and was cooled down to LN_2 temperature. The temperature was measured by a Si diode thermometer, and was stabilized better than 1 mK. Furthermore, the equilibrium pressure was measured simultaneously. It is known that the pressure isotherm bends clearly at the C-IC transition [3]. ¿From this bend, the fractional coverage of Kr lm was determined and the gas dosage was calibrated.

The resonance frequency of the quartz tuning fork was measured using a transmission circuit. In the circuit, the tuning fork was placed in series with a coaxial line connecting to a 50 Ω cw signal generator and a lock-in ampli er. The frequency of the signal generator was controlled in order to keep the in-phase signal zero, and was locked to the resonance frequency. The quadrature signal at this frequency was the resonance amplitude.

3. Results and discussion

3.1. Coverage dependence

Tuning fork experiments were carried out at 79.8 K with increasing the coverage of Kr lm monotonously. Figure 1 (a) shows the variation of the resonance frequency as a function of fractional coverage, together with that of 5 MHz QCM experiments at 85.8 K. The oscillation amplitude is about 1.4 nm in both experiments, and the oscillation direction is inclined at about 30° with respect to the a-axis of graphite. The vertical axis is normalized by the mass loading at the commensurate phase.

In the present experiments, the resonance frequency does not change greatly in the uid (F) phase and the C+F phase. When the lm enters the C phase, it starts to decrease. This decrease continues at the IC phase, and the ratio of decrease with respect to coverage at the IC phase is

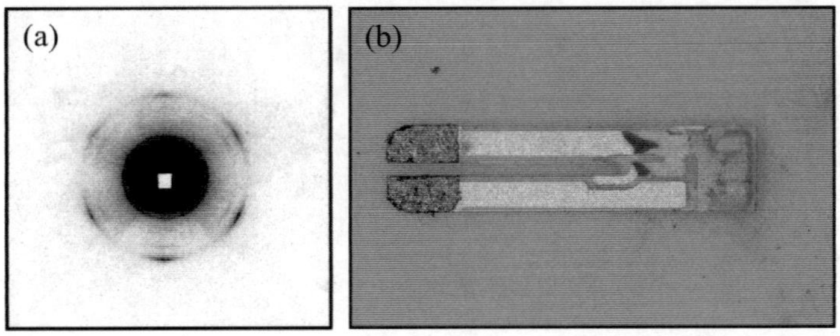

Figure 1. (a) Laue photograph of the graphite after exfoliation. (b) Photograph of the quartz tuning fork with graphite. The graphite is bonded on the tips of the Ag-plated arms.

larger than that of the C phase. The rst layer completion occurs around a coverage of 1.1 and the lm becomes bilayer in the higher coverage. Above a coverage of 1.4, the frequency becomes almost constant.

It is interesting to compare this behavior with that of 5 MHz QCM experiments. Since the inertial force is proportional to the square of the frequency, the force acting on lm of the present experiments is about four orders of magnitude smaller than that of 5 MHz QCM experiments. It was found that the observed behavior for the present experiments is similar to that of 5 MHz QCM experiments, although the frequency for 5 MHz QCM experiments decreases slightly at the C+F phase [4]. It should be noted that the decrease of the frequency at the C phase for both experiments is signi cantly smaller than the mass loading. This means that the lm of the C phase undergoes decoupling from the oscillating substrate under both experimental conditions.

Figure 1 (b) shows the variation of the resonance frequency in the vicinity of the C-IC transition. For the present experiments, the frequency decreases with increasing the coverage at the C phase. In the coexistence region between the C and IC phases, the decrease in frequency becomes small. Then, it is enhanced again at the IC phase. The coverage dependence of the coupled mass changes at the C-IC transition. The observed behavior is, however, di erent from a naive idea that the lm of the C phase does not slide easily because of the structural pinning [2]. We also compare the present results with that of 5 MHz QCM experiments. It was found that the observed behavior for the present experiments is similar to that of 5 MHz QCM experiments. The bend at the coexistence region seems to be somewhat clear compared with 5 MHz QCM experiments of 1.4 nm.

3.2. Sliding direction dependence

In the present experiments, we set two quartz tuning forks in the sample cell and measured them simultaneously. One of the tuning forks oscillates in the direction parallel to the a-axis of graphite, while the other in the direction inclined at about 30°. Figure 2 shows the variation of the resonance frequency for the two tuning forks. As seen in the gure, two sets of data almost coincide with each other. At present, we have no clear explanation that the direction dependence is very weak. There is, however, a possibility that the adatoms of lm on the substrate slide not along the oscillation direction but along the inclined direction which is the direction to the saddle point of the corrugation potential.

3.3. Comparison with ^{4}He films adsorbed on graphite

It is interesting to compare Kr/Gr with that of other noble gas lms adsorbed on graphite. Here, we make a comment on ^{4}He lms adsorbed on graphite (^{4}He/Gr). Tuning fork experiments for ^{4}He/Gr were carried out at 1.0 K, and the oscillation amplitude is 10 nm [6]. Figure 4 shows

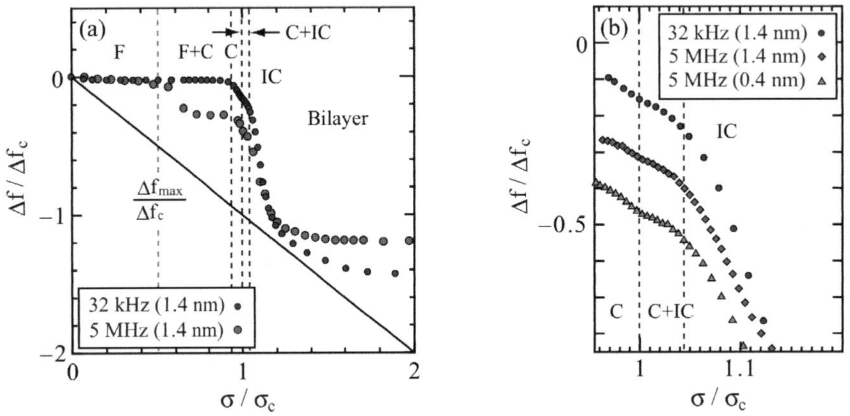

Figure 2. (a) Comparison of the coverage dependence between the present experiments at 79.8 K and 5 MHz AT-cut QCM experiments at 85.8 K [1]. The horizontal axis is the fractional coverage divided by the value of the commensurate phase (6.36 atoms/nm^2). The vertical axis is normalized by the mass loading at the commensurate phase. The estimated mass loading is represented by the solid line. C, F, and IC indicate the coverage region of the commensurate, the uid, and the incommensurate phases, respectively. These phases of Kr lm are determined on referring to Refs. [3] and [5]. (b) Expanded view in the vicinity of the commensurate-incommensurate (C-IC) transition.

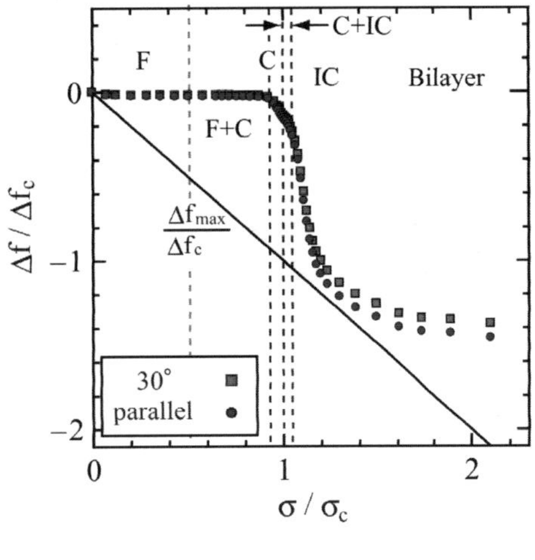

Figure 3. Comparison of the coverage dependence between two quartz tuning forks; One of the tuning forks oscillates in the direction parallel to the *a*-axis of graphite, while the other in the direction inclined at about 30°.

the comparison of the coverage dependence between Kr/Gr and ^{4}He/Gr. The horizontal axis is the coverage divided by the value of the rst layer completion σ_{1st}, and the vertical axis is normalized by the mass loading at σ_{1st}. For ^{4}He/Gr, the resonance frequency starts to decrease before the C phase, and continues to decrease with increasing coverage. At σ_{1st}, the decrease is close to the mass loading. It was found that in the monolayer region, ^{4}He/Gr does not slide easily compared with Kr/Gr.

We consider the inertial force acting on the lms and the corrugation potential for Kr/Gr and ^{4}He/Gr. It is expected that the lm undergoes slipping from the oscillating substrate when the inertial force F is large and the corrugation potential V is small, and the ratio of two values F/V may be a measure for decoupling. When the lm moves in concert with the oscillating

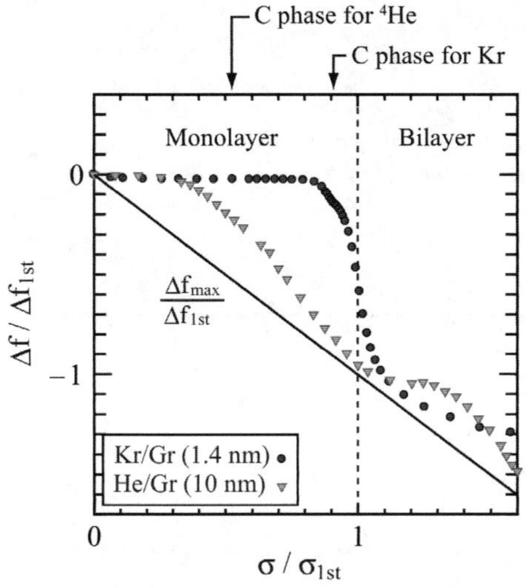

Figure 4. Comparison of the coverage dependence between Kr/Gr and ^{4}He/Gr. The horizontal axis is the coverage divided by the value of the rst layer completion σ_{1st}, and the vertical axis is normalized by the mass loading at σ_{1st}. The estimated mass loading is represented by the solid line. The arrows indicate the C phase for ^{4}He/Gr and Kr/Gr, respectively [7].

substrate, the amplitude of inertial force per adatom is expressed by $F = mA\omega^2$, where m is the mass of atom, A is the oscillation amplitude and ω is the angular frequency. Then, these values are shown in Table 1. As seen in the table, F/V for Kr/Gr is approximately equal in magnitude to that of ^{4}He/Gr, and we expect that they show a similar behavior. Contrary to this expectation, the coverage dependence for ^{4}He/Gr is clearly di erent from Kr/Gr: ^{4}He/Gr in the monolayer region does not slide easily compared with Kr/Gr. At present, the origin is an open question.

Table 1. Comparison between Kr/Gr and ^{4}He/Gr

	Atomic weight of adatom m (u)	Oscillation amplitude A (nm)	Corrugation potential^a V (K)	F/V (1/m)
Kr/Gr	83.80	1.4	76-126	4.5-7.5×10 ^{3}
^{4}He/Gr	4.00	10	24-43	4.5-8.1×10 ^{3}

^{a} From Ref. [8].
^{b} The resonance angular frequency ω is $2\pi\times$ 32 kHz.

4. Summary

We have developed QCM using a 32 kHz quartz tuning fork resonator, and have carried out experiments for Kr lms adsorbed on a single-crystalline graphite at LN_2 temperature. It was found that the coupled mass to the oscillating substrate does not depend on the oscillation direction with respect to the crystal axis of graphite. In addition, the coupled mass is signi cantly small until the lm enters the C phase. Then, it starts to increase slightly at the C phase, and it is enhanced when the lm enters the IC phase. The coverage dependence is di erent from a naive idea that lms of the C phase do not slide easily, and is similar to the previous experiments of 5 MHz AT-cut QCM. From the comparison between Kr/Gr and ^{4}He/Gr, we concluded Kr/Gr slides more easily than ^{4}He/Gr.

References

[1] Kobayashi H, Taniguchi J, Suzuki M, Miura K, and Arakawa I 2010 *J. Phys. Soc. Jpn.* **79** 014602
[2] Smith E D, Robbins M O, and Cieplak M 1996 *Phys. Rev.* B **54** 8252
[3] Hoja R and Marx R 1986 *Phys. Rev.* B **34** 7823
[4] Although the frequency decreases at the C+F phase in 5 MHz QCM experiments of Fig. 1 (a), this decrease depends on sample preparation. At present, it is not fully understood, but may be from the impurity pinning.
[5] Butler D M, Litzinger J A, Stewart G A, and Griffiths R B 1979 *Phys. Rev. Lett.* **42** 1289; Butler D M, Litzinger J A, and Stewart G A 1980 *ibid.* **44** 466
[6] Nihei F, Ideura K, Kobayashi H, Taniguchi J, and Suzuki M, to be appear in the proceedings of QFS2010.
[7] Greywall D S and Busch P A 1991 *Phys. Rev. Lett.* **67** 3535; Greywall D S 1993 *Phys. Rev.* B **47** 309
[8] Vidali G and Cole M W 1984 *Phys. Rev.* B **29** 6736

International Conference on Science of Friction 2010 (ICSF2010) IOP Publishing
Journal of Physics: Conference Series **258** (2010) 012019 doi:10.1088/1742-6596/258/1/012019

Development of a combined atomic force microscope with an AT-cut quartz resonator

Daisuke Inoue[1], Nariko Hosomi[1], Junko Taniguchi[1], Masaru Suzuki[1], Makoto Ishikawa[2], and Kouji Miura[2]

[1]Department of Engineering Science, University of Electro-Communications, Chofu,Tokyo 182-8585, Japan

[2]Department of Physics, Aichi University of Education, Kariya 448-8542, Japan

E-mail: dinoue@phys.uec.ac.jp

Abstract. To study the microscopic mechanism of energy dissipation due to sliding motion, we are developing a combined atomic force microscope (AFM) with a quartz crystal microbalance (QCM) resonator. In this system, the loading force is controlled by AFM, while the energy dissipation is measured by the change in Q value of the quartz crystal. At room temperature, we have checked the energy dissipation for several substrates as a function of loading force. The energy dissipation depends both on substrate material and on oscillation amplitude.

1. Introduction

Physical property of small contacting asperities is a subject of study attracting the interest of many researchers because of the importance to macroscopic and microscopic applications. Characterization of nanoscale contacting asperities have been performed by means of nanoindentation, atomic force microscopy (AFM). It is recognized that a probe tip-quartz crystal microbalance (QCM) has large potential as a sensitive probe of the physical properties [1, 2, 3]. In 1991, Sasaki et al. combined a scanning tunneling microscope (STM) with an AT-cut quartz resonator and reported a simultaneous image of STM surface topographies and a frequency shift of an AT-cut quartz resonator for gold thin films.

To study the microscopic mechanism of energy dissipation due to sliding motion, we have started to develop a combined atomic force microscope (AFM) with an AT-cut quartz crystal resonator which can operate even under low temperature condition. In this paper, we explain our combined AFM-QCM system, and also report preliminary results of the energy dissipation for several substrates as a function of loading force at room temperature.

2. The AFM-QCM system

Figure 1 shows a schematic diagram of our AFM-QCM combined system. The loading force is controlled by AFM, while the energy dissipation is measured by the change in Q value of the quartz crystal. The advantage of this system is to measure directly the the energy dissipation due to sliding motion by changing the magnitude of displacement in the nanometer scale and the loading force. A detail description of this system will be given below.

Journal of Physics: Conference Series **258** (2010) 012019 doi:10.1088/1742-6596/258/1/012019

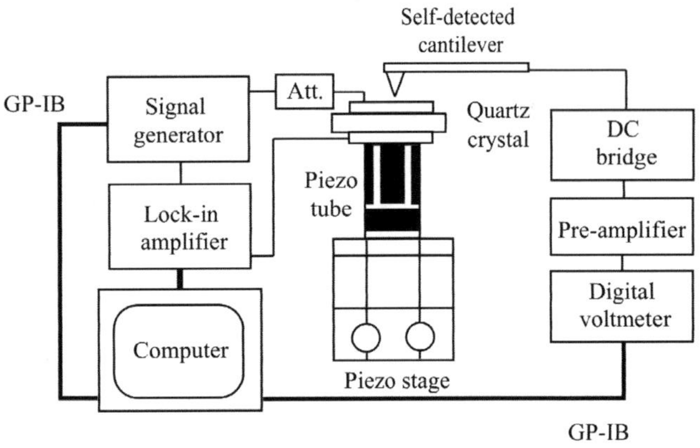

Figure 1. Schematic diagram of a AFM-QCM combined system.

2.1. AFM

For the AFM system, we use a Si self-detecting cantilever (NPX1CTP003, SII) of a spring constant of 4 N/m as a sensor of the loading force, which can operate easily even under low temperature condition [4]. The typical radius of the probe-tip is 20 nm. The loading force acting on the cantilever was measured by the change in piezoresistance of its joint. Its change $\Delta R/R$ is related to the loading force N as [5]

$$\frac{\Delta R}{R} = {}_L \frac{6LN}{(2a)^2 t^2} , \tag{1}$$

where ${}_L$ is the longitudinal piezoresistive coefficient for Si along the $\langle 001 \rangle$ direction, LN is the torque, and L is the length of the cantilever. Here, $2a$ and t is the width and the thickness of the joint which is a part of the cantilever deformed by the loading force. In our case, $a \simeq 2$ μm, $t \simeq 4$ μm and $L \simeq 400$ μm. Then, the loading force of 1 nN corresponds to $\Delta R/R$ 11 ppm.

The change in resistance is measured using a DC bridge circuit. The reference resistance is also fabricated in the basement of cantilever. The resistance of the cantilever and reference is 600-700 Ω, and the excitation voltage is 1.0 V. The out-of-balance signal is amplified by the preamplifier (LI-75A, NF Corporation). Then, we can easily measured the loading force less than 1 nN.

The position of the substrate is driven by a linear positioner (ANPz101, atocube systems) for the coarse adjustment and a piezo tube actuator for the fine adjustment. The sensitivity of the piezo tube actuator is 1.86 nm/V, and the applied voltage is swept from -100 V to 100 V.

2.2. QCM

The energy dissipation caused by the sliding motion between the probe-tip and the substrate attached on the crystal is related to $\Delta(1/Q)$ of the quartz crystal as

$$\frac{\dot{E}}{E} = \omega_0 \Delta \quad \frac{1}{Q} \quad , \tag{2}$$

where E is the oscillation energy of the crystal, and $\omega_0 = 2 \ f_0$ is the resonance angular frequency. In this system, the resonance frequency and amplitude are measured using a transmission circuit. The crystal is placed in series with a coaxial line connecting to a 50 Ω signal generator and an RF lock-in amplifier. The frequency of the signal generator is controlled in order to keep the in-phase signals zero, and is locked to the resonance frequency. The quadrature signal at this frequency is the resonance amplitude. Here, $\Delta(1/Q)$ is obtained as

$$\Delta \quad \frac{1}{Q} \quad = -\frac{1}{Q} \cdot \frac{\Delta A_q}{A_q} \quad , \tag{3}$$

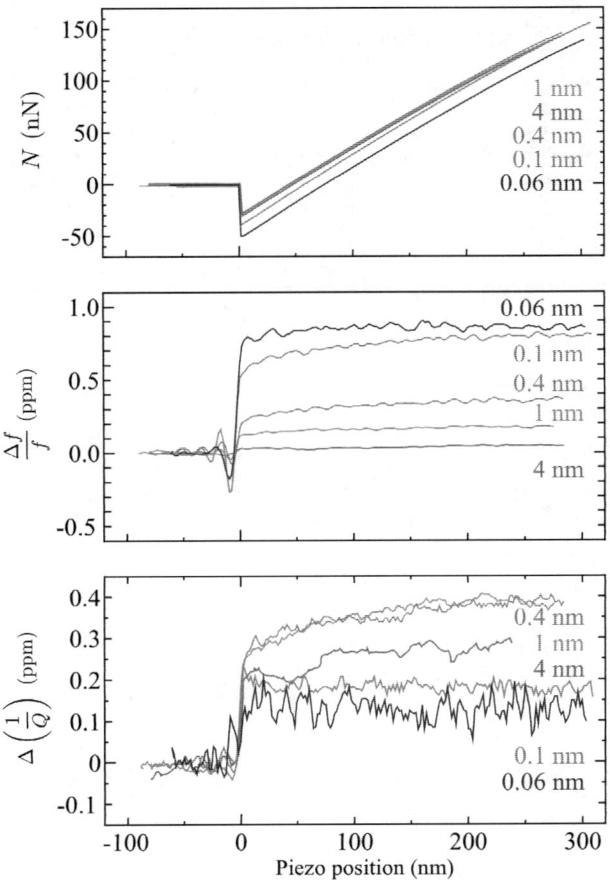

Figure 2. Piezo position vs (a) the loading force, (b) the resonance frequency, and (c) the Q value when retracting the cantilever from HOPG surface.

where A_q is the magnitude of the quadrature signal. The oscillation amplitude of the crystal is controlled by the alternating input voltage, and is estimated from the magnitude of the quadrature signal and the oscillating area of the crystal with the substrate. In the present experiments, the input voltage is in the range of 50 to 0.5 mV_{pp}, which corresponds typically to the oscillation amplitude of 4 to 0.06 nm [6].

A small thin substrate (1 1 mm^2) is attached on the electrode of an AT-cut quartz crystal with the resonance frequency of 3.2 MHz. After attaching the substrate, the Q value remains on the order of 10^4.

3. Energy Dissipation vs Loading Force

We have checked the energy dissipation for several substrates at room temperature as a function of loading force. It was found that the energy dissipation depends both on substrate materials and on oscillation amplitude.

First, we briefly explain the energy dissipation on the assumption of some conditions. In case when the sliding friction between the probe-tip and the substrate is described as the Amontons-Coulomb law $F = \mu' N$, where N is the loading force, $\Delta(1/Q)$ is obtained as

$$\Delta \ \frac{1}{Q} \ = \frac{8}{\ ^2} \cdot \frac{\mu' N}{\omega_0 \rho v A S} \ , \tag{4}$$

where ρ and v are the density and the sound velocity of the quartz crystal, and A is the oscillation amplitude. Here, S is the oscillating area of the crystal. As seen from this equation, $\Delta(1/Q)$

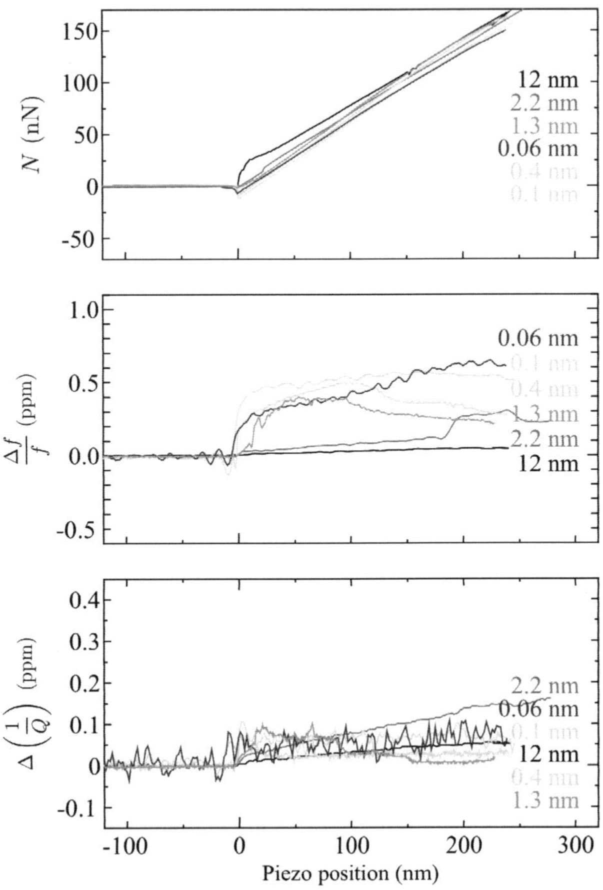

Figure 3. Piezo position vs (a) the loading force, (b) the resonance frequency, and (c) the Q value when retracting the cantilever from mica surface.

is in inverse proportion to the amplitude since the oscillation energy increases proportional to the square of amplitude. In contrast, $\Delta(1/Q)$ does not depend on amplitude when the sliding friction is a viscous friction.

Figures 2 and 3 show the piezo position dependence for various oscillation amplitudes when retracting the cantilever from HOPG and mica surfaces.

First, we examine the piezo position dependence for HOPG surface. The top panel is the loading force as a function of piezo position. When retracting the cantilever, the loading force decreases almost linearly and the cantilever shows a single sharp jump-out. The attractive force at the jump-out depends on oscillation amplitude. When the amplitude becomes small, this force increases drastically. The middle panel is the change in the frequency $\Delta f/f$ as a function of piezo position. It was found that the frequency does not depend strongly on loading force, and shows a stepwise change at the contact between the probe-tip and the substrate surface. In addition, the frequency clearly shows the oscillation amplitude dependence. In the experimental range of amplitude, it decreases monotonously from 0.8 to 0.05 ppm when the amplitude increases from 0.06 to 4 nm. The bottom panel is the change in Q value as a function of piezo position. ΔQ^{-1} shows an interesting behavior. On the whole, ΔQ^{-1} takes the maximum at around an oscillation amplitude of 0.4 nm. At this amplitude, ΔQ^{-1} shows a stepwise change at the contact and increases gradually with increasing loading force. When the amplitude decreases from this value, ΔQ^{-1} becomes small and does not depend on loading force. On the other hand, when the amplitude increases to 4 nm, ΔQ^{-1} becomes small again. This behavior is reproducible regardless of the position of the surface.

As mentioned above, when the sliding friction obeys the Amontons-Coulomb law, ΔQ^{-1} is in inverse proportion to the amplitude. On the other hand, in the case of a viscous friction, ΔQ^{-1} does not depend on amplitude. In contrary to these expectations, ΔQ^{-1} increases with increasing the amplitude and takes the maximum at 0.4 nm. On the other hand, $\Delta f/f$ decreases monotonously with increasing the amplitude. This means that the sliding motion has a typical length in which the energy dissipation becomes large. At present, the mechanism for the typical length is an open question, but the typical length may be compared with the lattice distance.

Next, we make a short comment on mica surface. Concerning the loading force, the attractive force at the jump-out is small. Furthermore, ΔQ^{-1} is significant small. In a large oscillation amplitude, we observed that $\Delta f/f$ sometimes changes rapidly during piezo position scan.

4. Summary

We are developing AFM combined with an AT-cut quartz resonator. In this system, the loading force is controlled by AFM, while the energy dissipation is measured by the change in Q value of the crystal. At room temperature, we measured the energy dissipation for several substrates as a function of loading force and oscillation amplitude. For HOPG surface, the energy dissipation does not depend strongly on loading force up to about 150 nN. Concerning the amplitude dependence, ΔQ^{-1} increases with increasing the amplitude and takes the maximum, while the $\Delta f/f$ decreases monotonously. This means that the sliding motion has a typical length in which the energy dissipation becomes large. At present, the mechanism for the typical length is an open question.

References

[1] Sasaki A, Katsumata A, Iwata F, and Aoyama H 1993 *Appl. Phys. Lett.* **64** 124
[2] Borovsky B, Krim J, Syed Asif S A, and Wahe K J 2001 *J. Appl. Phys.* **90** 6391
[3] Berg S and Johannsmann D, 2003 *Phys. Rev. Lett.* **91** 145505
[4] Ishiakwa M, R. Harada, Sasaki N, and Miura K 2008 *Appl. Phys. Lett.* **93** 083122
[5] Rossel C, Bauer P, Zech D, Hofer J, Willemin M, and Keller H 1996 *J. Appl. Phys.* **79** 8166
[6] The oscillation amplitude is in proportion to both the alternating input voltage and the Q value of the crystal.

AUTHOR INDEX